BRITISH
MEDICAL BULLETIN

VOLUME FORTY-FIVE
1989

.

CHURCHILL LIVINGSTONE
EDINBURGH, LONDON, MELBOURNE AND NEW YORK

CHURCHILL LIVINGSTONE
Medical Division of Longman Group UK Limited

Distributed in the United States of America by Churchill
Livingstone Inc., 1560 Broadway, New York, NY 10036, and
by associated companies, branches and representatives
throughout the world.

ISSN 0007-1420
ISBN 0 443 04199 7

BRITISH MEDICAL BULLETIN

British Medical Bulletin is published four times each year, in January, April, July and October.

Subscriptions and single-copy orders should be sent to: Longman Group UK Ltd, Subscriptions Department, Fourth Avenue, Harlow, Essex CM19 5AA.

Subscription rates for 1989 are: £70 (UK) or £87.50/$157.50 (overseas)

Single copies will be available at £25.00 (UK) or £31.50/$49.25 (overseas)

NEXT ISSUE

October 1989

CHURCHILL LIVINGSTONE, Medical Division of Longman Group U.K. Limited. Typeset and printed by H Charlesworth & Co Ltd, Huddersfield

Distributed in the United States of America by Churchill Livingstone Inc., 1560 Broadway, New York, NY 10036, and by associated companies, branches and representatives throughout the world. This journal is indexed, abstracted and/or publishd online in the following media: Current Contents, Scientific Serials Review, Excerpta Medica, USSR Academy of Science, Biological Abstracts, UMI (Microform), BRS Colleague (full text), Index Medicus, BIOSIS, NMLUIS, Adonis

© The British Council 1989

ISSN 0007-1420   ISBN 0 443 04199 7

Molecular Genetics of Muscle Disease— Duchenne and Other Dystrophies

Scientific Editors: *A J Buller, J Goodfellow, J M Newsom-Davies*

Professor Buller chaired the committee which included Dr J Goodfellow and Professor J M Newsom-Davies that planned this number of the *British Medical Bulletin.* We are grateful to them for their help, and for acting as Scientific Editors for the number.

British Medical Bulletin is published by Churchill Livingstone for The British Council, 10 Spring Gardens, London SW1A 2BN

British Medical Bulletin (1989) Vol. 45, No. 3, pp. 605–607
© The British Council 1989

Introduction

A J Buller
Steeple Aston, Oxon, UK

By the time this volume of the British Medical Bulletin makes its appearance it will have been something over nine years since the publication of Volume 36 Number 2 of the old format BMB entitled 'The Muscular Dystrophies'. It is true that the title of the present volume is 'Molecular Genetics of Muscle Disease' but the subtitle 'Duchenne and other dystrophies' clearly marks it as a 'follow on' volume.

Readers who are interested in the manner in which scientific ideas evolve or regress would be well advised to return to the former issue and browse through it before venturing on to what follows this brief introduction to the present volume.

In the May 1980 BMB only one contribution referred to the impact that recombinant DNA technology might make on the understanding of Duchenne muscular dystrophy. Whilst Dr K W Jones advocated screening DNA with probes derived from myogenic cell mRNAs which in fact was not to provide the way forward, his article clearly set out the potential value of the new recombinant DNA techniques. He concluded 'In view of the foregoing sketch of the methods and prospects for genetic engineering in muscle research it seems fair to conclude by asking the question: what are the prospects for research into muscle cell differentiation and pathogenesis without the application of these revolutionary new approaches?' Ken Jones' paper received no mention in Sir Andrew Huxley's summarizing contribution which was entitled 'Future prospects'. It would seem that both researchers and those concerned with the support of research into the muscular dystrophies were ill-prepared for what was about to happen! Within two years of the publication of the 1980 BMB Dr Kay Davies and her colleagues had published their findings using X chromosome DNA probes which localized the genetic defect causing Duchenne muscular dystrophy to the region Xp21.

There are two points that I would wish to make concerning these germinal studies which have rightly been claimed to have 'put the DMD gene on the map'. The first is that the work would not have been possible without an enormous investment in 'basic' science. In the UK the investment over many years by the Medical

0007–1420/89/0045–0605/$10.00

Research Council (MRC) in molecular genetics, not least at the Laboratory of Molecular Biology, Cambridge, had frequently been criticised as being 'unrelated to the practice of medicine'. Without the techniques developed during this period Kay Davies' work and much of what is reported in this BMB would have been impossible. The second point is that at the time of the 1982 paper Kay Davies was working at St. Mary's Hospital (London) in Professor R Williamson's department. That laboratory was receiving substantial funding from a medical charity, the Cystic Fibrosis Research Trust.

The latter observation underlines two further points. The remarkable progress in our understanding of the Duchenne gene has not occurred in isolation, neither is it unique. The 'New Genetics', masterfully underwritten in the UK by Sir David Weatherall, involves all the genetically determined diseases. The X linked diseases inevitably had a head start because of their pattern of inheritance, but many autosomal diseases followed, or are following, closely behind. The muscular dystrophies are not the only conditions that have benefited from the earlier investment in 'basic' science. Without that investment the present mushrooming of disease-orientated research could not have occurred.

It is also pertinent to note that the contribution from the private sector, as opposed to that from the public purse, has been dominant in sustaining the research effort needed to further understanding of the genetically determined diseases. The financial contribution made towards the cost of the work on the Duchenne gene by the Muscular Dystrophy charities in this country, in the USA and elsewhere has been crucial in reaching the present state of knowledge. Neither have the charities been overprotective of their investments. In the same manner that one of the earliest probes used to localize the Duchenne gene originated from a laboratory receiving funding from the Cystic Fibrosis Research Trust, so other inherited diseases have been illumined by observations made in laboratories supported by money provided by the Muscular Dystrophy charities. Perhaps this is as it should be. If the Medical Research Council continues to use the funds provided from the public purse to support basic research which, in the nature of things, has little appeal (but much interest) to the layman it is probable that charitable giving will cover the cost of many of the applications of this knowledge to specific conditions and diseases. If the Research Council does not undertake the longer term basic research it will not be done.

The contributions to this volume clearly illustrate the international effort which brought about the current level of understanding of the Duchenne gene. In the past it has not been uncommon for knowledge to have been built up as a result of contributions from many countries. What has been remarkable in the work described in this BMB is the international collaboration and goodwill that has pervaded the research. The exchange of DNA probes, biopsy material and now antibodies has been both generous and remarkable.

The present volume marks that stage which was so vividly set out by Andrew Huxley in 1980 'Nevertheless, the discovery of the essential defect in Duchenne muscular dystrophy, whenever it may come, will be a tremendous advance. It may lead to a direct cure, or at least to effective control of the clinical manifestations, as was the case for phenylketonuria; it will certainly lead to a deeper understanding of the disease process and hence to possibilities of treatment related to one of the steps in that process; and it will almost inevitably lead to more effective carrier detection and antenatal diagnosis'.

Effective carrier detection and antenatal diagnosis are now available as predicted by Andrew Huxley. But as yet there is no cure for any of the muscular dystrophies. Such progress must wait on further basic research. One hopes that such work will not be frustrated by inappropriate legislation. However there must be the possibility that effective treatments for these conditions will antedate cures, and the opportunity for a determined effort to find such treatments now exists and must be taken. In the meantime the good physician is left to provide expert management of patients with neuromuscular disease so that their quality of life is optimized and the public is able to appreciate the very real qualities and talents possessed by many of those who must live their lives as 'dystrophics'. For them I hope that the next BMB devoted to the muscular dystrophies will provide some of the answers they hope for. Now read on.

British Medical Bulletin (1989) Vol. 45, No. 3, pp. 608–629
© The British Council 1989

The control of muscle gene expression: A review of molecular studies on the production and processing of primary transcripts

M E Buckingham
Pasteur Institute, Paris, France

Many muscle proteins exist as multiple isoforms. This diversity is generated by the presence of multigene families and by alternative splicing of individual genes. Examples are given of different modes of alternative splicing undergone by the primary transcripts of muscle genes, and preliminary studies on the mechanism are mentioned. The chromosomal organization of muscle genes is discussed briefly. Studies on their transcriptional regulation are reviewed first in terms of cis-acting sequences in the proximal promoter region and elsewhere in the vicinity of the gene, which are necessary for its expression, and, secondly, in terms of trans-acting transcriptional factors which interact with such sequences. Molecular regulation of splicing and of transcription is discussed mainly with reference to the muscle genes of mammals.

GENERATION OF ISOFORM DIVERSITY AMONG MUSCLE PROTEINS

Many muscle proteins exist as multiple isoforms, characteristic of different types of muscle, and indeed frequently present in non-

0007–1420/89/0045–0608/$10.00

muscle tissues too (for review *see* Ref. 1). This is particularly striking for the contractile proteins. Actin, the major structural protein of the thin filament of the muscle sarcomere, for example, is present as six different isoforms in higher vertebrates, two found in striated muscles, (α-cardiac and α-skeletal), two in smooth muscles (α- and γ-smooth), and two in all non-muscle cells (β- and γ-cytoplasmic). Myosin heavy chain molecules show even more diversity since not only are different isoforms present in different adult muscle types, and in non-muscle cells, but there are also distinct developmental myosin heavy chains present in fetal and neonatal muscle. This phenomenon is not confined to sarcomeric proteins. The acetylcholine receptor, a multimeric protein located in the muscle membrane at the synaptic junction with the nerve also has fetal and adult subunits. Creatine phosphokinase, a major enzyme in muscle metabolism (M-CPK), is present in some muscle and in non-muscle cells as a different isozyme (B-CPK).

Why there should be such isoform diversity is not always evident. In the case of the myosin heavy chains, for example, different isoforms have different ATPase activities which are presumably adapted to the physiological context of the muscle. However the *raison d'être* for the different myosin alkali light chains is not clear and this is true, too, for the isoforms of actin. Cardiac and skeletal actins are remarkably conserved, with only 4 substitutions in 376 amino acid residues, while the difference between non-muscle and muscle actins is less than 8%.[2]

Skeletal and cardiac actins are to some extent interchangeable in vivo. In a mutant mouse, the presence of relatively lower levels of cardiac actin and higher levels of skeletal actin does not seem to impair the viability of the mouse.[3] However, these minor differences between isoforms are conserved between mammalian and avian species suggesting that they are significant for the functional fine-tuning of the muscle. The question of isoform function can best be addressed by manipulating the sequence in vivo. Such experiments are being carried out for the actins in *Drosophila* for example,[4] and should be feasible in the future in vertebrates. The development of homologous recombination techniques in mice[5] should permit the elimination or modification of a gene in the germ line so that the effects at the protein level can be studied in the developing and mature muscle of the mouse.

How such diversity of muscle isoforms is generated is an important consideration in any discussion of muscle gene regulation. Essentially two strategies can be distinguished. Muscle genes are organized into multigene families where a distinct gene encodes each isoform, or alternatively, the same gene encodes two or more isoforms which are generated by different combinations of exons during splicing of the pre-messenger RNA in the nucleus. In the actin multigene family, isoform diversity is not generated by differential splicing: different actins are encoded by different genes. In the case of the myosin heavy chains, the strategy adopted varies with the species (see Ref. 1). In higher vertebrates the diversity of muscle isoforms is reflected in a corresponding number of genes, whereas in the invertebrate, *Drosophila*, myosin heavy chains present in muscle are derived from differential splicing of the same gene. For other muscle proteins, both strategies may be used in the same species. Thus in mammals there are two α-tropomyosin genes, each of which encodes several tropomyosin isoforms expressed in different muscle and non-muscle tissues. In the case of the myosin alkali light chains, a single gene may encode a single protein, as seen for the cardiac isoforms, or the gene may encode more than one protein, as for the fast skeletal muscle isoforms. In conclusion all combinations are found, for the same protein family between species, and for different families, within the same species. The generation of isoform diversity among the contractile proteins provides a good example of what François Jacob has described as 'molecular tinkering'.[6]

Alternative splicing: more than one isoform from a single gene

This mode of generating isoform diversity is relatively frequent in muscle, and indeed has the potential advantage of maintaining part of the protein sequence strictly identical while introducing variation by the splicing of different exons in other parts of the sequence. In the case of the structural proteins of muscle which are inserted with strict stoichiometry into the sarcomere, the maintenance of invariant regions may be important. A number of different types of alternative splicing can be distinguished formally as follows (for review see Ref. 7):

(1) transcription is initiated at different promoters resulting in different 5' exons in the messenger RNA and possibly, depending on the organization of the 5' non-coding mRNA sequence, in different $-NH_2$ terminal sequences in the protein;

(2) transcription terminates differently because of more than one

site of polyadenylation resulting in different 3' exons, and hence, again depending on the 3' non-coding sequence of the gene, in different -COOH terminal protein sequences;

(3) different internal sequences are present either because of the alternative splicing of one or more different exons, or because in some situations an intron is spliced in as an exon giving an additional piece of sequence, or because of alternative splice sites within a single exon which remove a portion of the coding sequence.

Most of these possibilities are observed in muscle gene splicing, frequently in combination. Thus the two myosin alkali light chains of fast skeletal muscle in mammals and birds are the products of a single gene. Transcription is initiated from two different sites and differential splicing of internal exons subsequently takes place to generate two different 5' exons for each of the light chains (see Fig. 1). The rest of the primary transcripts are spliced identically maintaining the -COOH end of the isoforms invariant.[8] Internal alternative splicing, combined with the use of different sites of polyadenylation at the 3' end of the gene is observed for the tropomyosins (Fig. 2). Thus two β-tropomyosins, one expressed in smooth muscle and fibroblasts and one in skeletal muscle are generated from the same gene. Each is derived from 9 exons, but

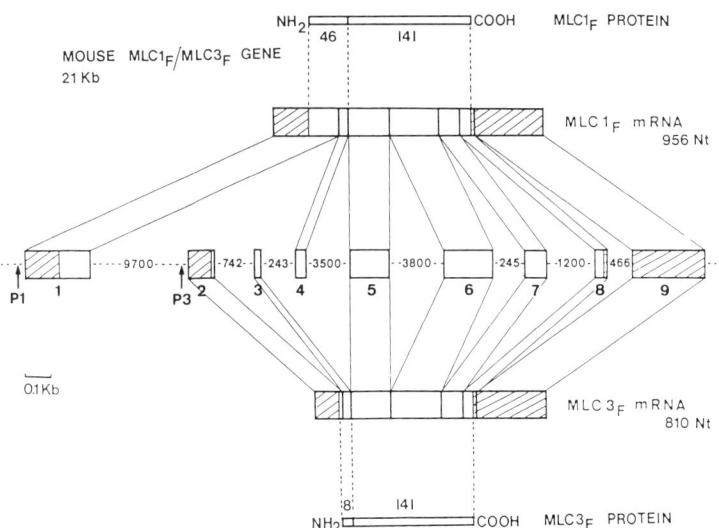

Fig. 1 Alternative splicing of the myosin alkali light chain gene expressed in adult fast skeletal muscle.[8]

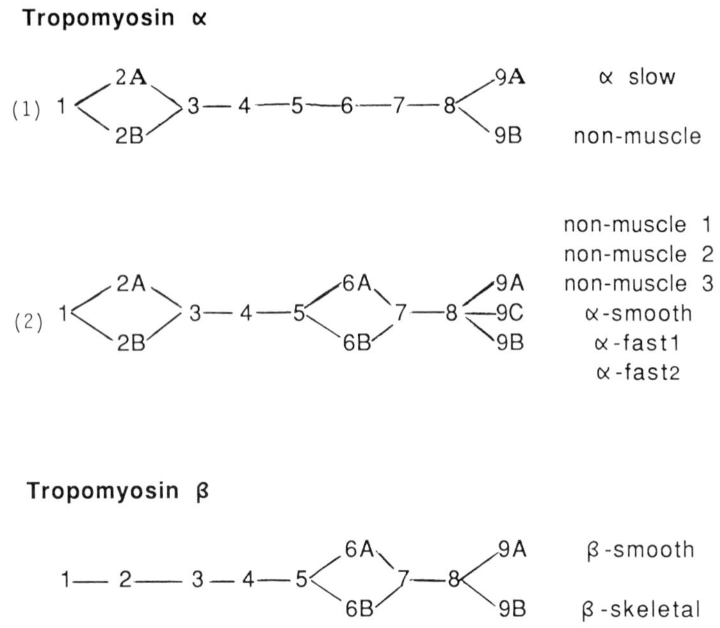

Tropomyosin α

Tropomyosin β

Fig. 2 Alternative splicing of mammalian tropomyosin genes.[9-11]

the internal exon 6 and terminal exon 9 differ.[9] Still more complex is the case of the α-tropomyosins which again are generated from 9 exons. In mammals there are two genes. One generates a slow skeletal muscle and a non-muscle isoform by use of alternative exons at position 2, 6 and 9.[10] The second gene, again by variation of internal exons 2 and 6 and the terminal exon 9, gives rise to at least six different transcripts, each encoding a different isoform expressed specifically in different non-muscle, smooth, and striated (cardiac and skeletal) muscle cells.[11] Apart from one exon (2a) which is only found in smooth muscle tropomyosin α, it is difficult to interpret the splicing pattern in terms of tissue specific isoform function. It is interesting from an evolutionary point of view that in the invertebrate *Drosophila*, tropomyosin genes are present which show some of these splicing features too.[7] In the case of the myosin alkali light chains, too, the generation of splice patterns can be traced from *Drosophila* to mammalian non muscle, to muscle, with the provision of isoforms for increasingly specialized muscle types.[12,12a]

The muscle gene described in mammals which presents the

greatest complexity of alternative splicing is the troponin T gene expressed in fast skeletal muscle.[13] By different combinations of internal exons, two near the 3' end of the gene and 5 near the 5' end as many as 64 TnT messenger transcripts are generated. Whether there are as many different isoforms is not known and what the function of such a striking degree of variability may be is totally unclear at present. Other troponin T genes encoding slow skeletal[14] or cardiac[15] troponins also demonstrate alternative splicing but of a different and much more restricted type in each case. The troponin T gene family therefore differs from that of the tropomyosins where a similar alternative splicing strategy is conserved between genes. Alternative splicing is not only undergone by genes encoding sarcomeric proteins. It has been demonstrated that the neural cell adhesion molecule (N-Cam) present on the muscle cell surface contains a muscle specific sequence in the extracellular domain, as a result of alternative splicing of internal exons.[16]

Structural studies have established the different exon combinations employed, but the question of how alternative splicing is regulated is only beginning to be addressed. It is clear that it takes place in a tissue specific manner, in different muscles and in muscle and non-muscle tissues. In certain cases it is also developmentally regulated.[8,13] By making mini-gene constructions in vectors which can be transfected into cells, it is possible to ask questions about the sequences which are necessary for a splicing event. Thus it has been demonstrated that minigenes containing only a limited number of exons retain their capacity for alternative splicing,[7] indicating that this is determined by *cis*-acting sequences. In the case of the cardiac troponin T gene, for example, deletion analysis has limited *cis*-elements required for alternative splicing to three small regions of the primary transcript.[15] In these experiments tissue specificity was not retained and alternative splice products were seen in a variety of non-muscle cultured cells as well as in muscle cells, although transcripts of the cardiac troponin T gene are normally only present in muscle. This would suggest that the splicing mechanism which recognizes the sequence elements as splicing signals is not tissue specific, and that other factors such as the tertiary conformation of the complete pre-messenger RNA transcript may be important. In contrast mini-gene constructions of the troponin T gene which gives rise to multiple transcripts in skeletal muscle, are spliced with some degree of tissue specificity.[13] Splicing takes place in non-muscle cells, but the splice signals for the alternative exons are not recognized and these are only included in the messenger RNA in muscle cells. The authors therefore conclude that factors required for correct alternative splicing are induced during myogenesis.[13] The nature of such trans-acting splicing factors remains unknown. In fact, in order that the splicing process can be investigated further it will be essential to develop in vitro systems with cell extracts from different types of muscle cells. At present the only functional in vitro splicing system to be described is one derived from non-muscle HeLa cells[17] which does not permit study of tissue specific alternative splicing.

Chromosomal organization of muscle genes

Alternative splicing is a mechanism which contributes to the isoform diversity seen for certain muscle proteins, but, in almost all cases, where multiple isoforms exist these are encoded by several genes. The size of the multigene family is variable; for the myosin heavy chains, for example, in mammals, there are probably at least twelve different genes, whereas the number of different tropomyosin genes identified to date is less. The chromosomal organization of muscle genes is a potentially important factor in determining how their transcription is regulated. Are genes expressed in the same muscle phenotype adjacent to each other, and are members of the same multigene family linked? The answer to the first question is negative. In general there is no evidence for linkage, either physical or genetic between genes encoding different proteins which are co-expressed, in the same muscle[18] (Fig. 3). This therefore necessarily implies that regulatory mechanisms which co-ordinate the expression of a muscle phenotype operate in 'trans'. Linkage between genes of the same multigene family varies with the species; thus actin genes in several invertebrate and unicellular organisms are physically adjacent to each other, whereas in mammals and birds they are dispersed on different chromosomes.[18] Such linkage may reflect a relatively recent gene duplication even in a species, rather than any functional requirement. In mammals the members of most multigene families which have been examined are not linked. The myosin alkali light chain genes, for example, map to different mouse chromosomes. In the case of the genes encoding the subunits of the acetylcholine receptor, two are adjacent to each other, δ and γ, while the genes for α and β are elsewhere.[19] Again this linkage may simply reflect

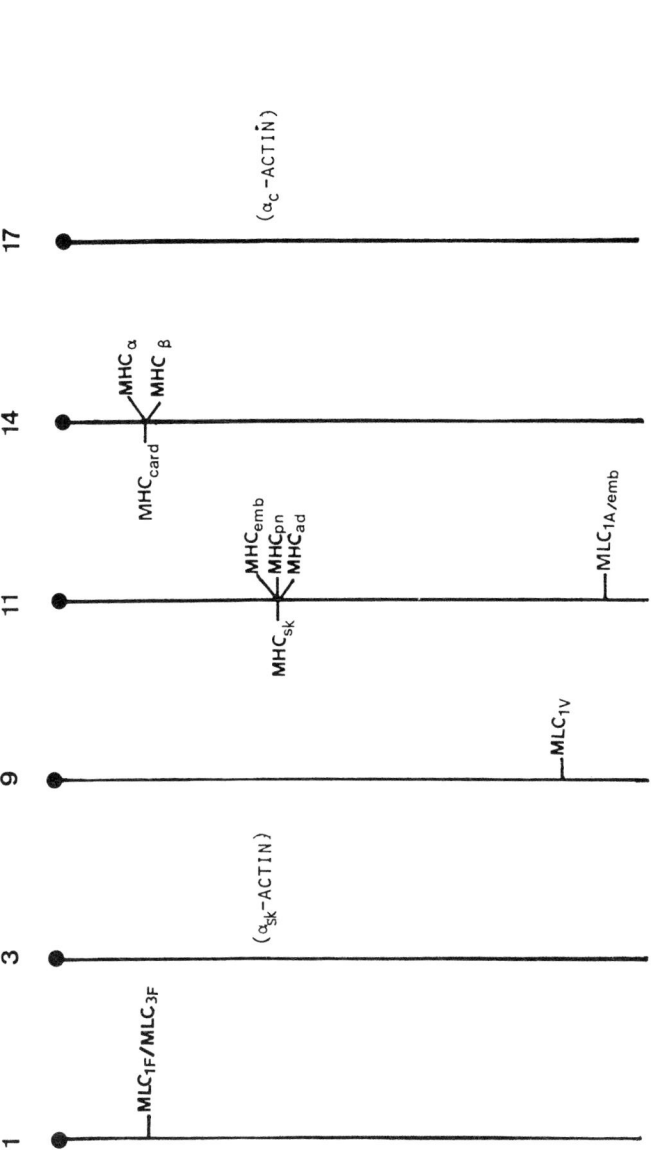

Fig. 3 Chromosomal organization of actin and myosin genes in the mouse. This illustrates the point that genes expressed in the same muscle phenotype are not linked,[18,20] for example:

MLC1$_F$/MLC3$_F$: α_{sk}-actin: MHC$_{ad}$—adult skeletal muscle

MLC1$_V$: α_c-actin: MHC$_\alpha$—adult ventricular muscle

MLC1$_F$: MLC1$_A$: α_{sk}-actin: α_c-actin: MHC$_{pn}$: MHC$_{emb}$—fetal skeletal muscle.

the evolutionary history of these genes. The most striking example of linkage among muscle genes of the same family is that of the myosin heavy chain genes. In the mouse genome there are two clusters on different chromosomes.[20] The two cardiac genes, α and β, are located in one cluster within a few kilobases of each other. The myosin heavy chain genes expressed in adult fast, perinatal, and fetal skeletal muscle are grouped together at a second chromosomal locus. There is no direct evidence to suggest that this linkage reflects a regulatory constraint. However, the genes within these clusters are developmentally regulated and display a mode of sequential expression in the heart or in skeletal muscle which is reminiscent of the α and β globin gene clusters where there is evidence for regulatory sequences which act in *cis* over long distances on genes in the cluster (*see* Ref. 20).

Regulatory sequences associated with muscle genes

At present, progress in understanding how muscle genes are regulated at the transcriptional level has principally been made where sequences in the immediate vicinity of the genes are concerned. To address the question of whether the regulatory signals necessary for the expression of a gene are present in a particular DNA sequence, this sequence is placed in front of an indicator gene in an appropriate vector and transfected into cells in culture. Detection of the product of the indicator gene suggests that the sequence promotes transcription. Regions of a few hundred kilobases immediately proximal to the 5′ ends of a number of muscle genes have been shown to contain the information necessary for expression of these genes, in cultured muscle cells. The next question is whether regulatory sequences are present which confer muscle specificity, that is whether the indicator gene is expressed only in muscle cells and not in non-muscle cells, and furthermore whether this expression is restricted to certain types of differentiated muscle cells in accordance with the expression pattern of the endogenous gene (Fig. 4). One of the best examples of a 5′ proximal sequence which promotes a high level of expression only in differentiated muscle fibres is that of the gene encoding the α-subunit of the acetylcholine receptor.[21] Other proximal promoter sequences, such as that from the gene for α-cardiac actin are most active in differentiated myotubes when the endogenous cardiac actin gene is expressed, but show some degree of 'leakiness' in non muscle cells.[21] Not many experiments have

myoblasts   **myotubes**

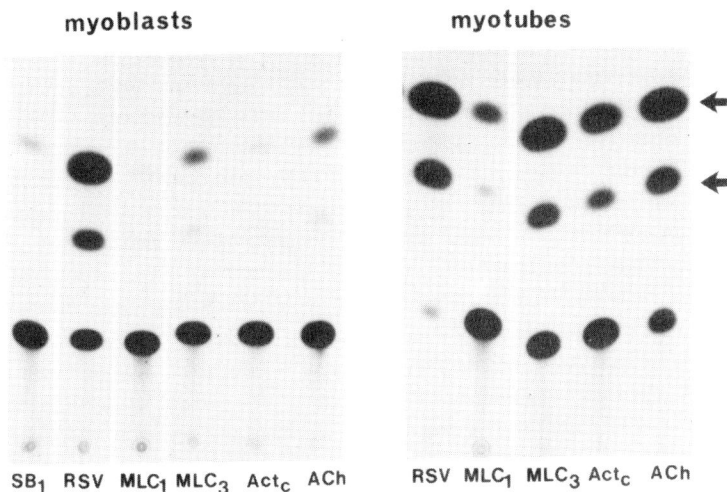

SB$_1$ RSV MLC$_1$ MLC$_3$ Act$_c$ ACh   RSV MLC$_1$ MLC$_3$ Act$_c$ ACh

Fig. 4 Transfection experiments with primary muscle cultures to test muscle regulatory sequences. The indicator gene is chloramphenicol acetyl transferase (CAT) and enzyme activity is indicated by conversion of the substrate chloramphenicol to its acetylated forms (←). The promoter constructions used were as follows: SB$_1$, negative control without promoter; RSV, positive control with viral promoter; MLC1 and MLC3, the 5′ upstream promoter regions of the mouse myosin alkali light chain gene of fast skeletal muscle; Act$_c$, the 5′ upstream promoter region of the mouse cardiac actin gene; ACh, the 5′ upstream region of the chick α-subunit acetylcholine receptor gene. All the muscle sequences show enhanced activity in myotubes (*see* Ref. 21).

been reported where the activity of sequences in different types of muscle cell has been compared. However the 5′ proximal regions of some myosin genes, which in vivo are expressed only in the heart promote expression when transfected into differentiated skeletal muscle cells as well as cardiocytes.[22] In one recent series of experiments a sequence in the proximal promoter region of the cardiac troponin gene was shown to be necessary for activity in embryonic cardiac cells, but not in skeletal muscle cells where the proximal promoter is also active.[23]

In 5′ proximal promoter regions of muscle genes it is clear that multiple regulatory elements are present. This is demonstrated by deletion analysis and by mutagenesis. These regions may exert a negative or a positive effect, preventing transcription in non-muscle cells and myoblasts as well as promoting it in differentiated muscle cells, as demonstrated by the mammalian embryonic myosin heavy chain[24] creatine phosphokinase[25] and skeletal actin[26] gene promoters. One of the best defined negative regulatory sequences is in the proximal promoter region of a cardiac atrial myosin light chain gene (MLC2A).[22] However most of the regulatory regions which have been defined, operate positively as activators of muscle gene expression. A large number of different sequences seem to act in this way, in addition to the classical TATA type sequence at about -30 nucleotides from the transcription initiation site. One of the best characterized of these positive elements is the CC(A + T rich)$_6$GG motif first described in the human cardiac actin gene promoter.[27,28] Four such sequences termed CArG boxes are present in this promoter (see Fig. 5). The first one is located at about -100, in the region where the classical CAAT box is found in many eukaryotic gene promoters. It has been shown that the first two CArG box sequences in this promoter act in a mutually dependent way to promote transcription in muscle cells. The first CArG box has most effect on transcription but can be replaced by the second CArG sequence, although with reduced activity. The distance between the two CArG boxes does not seem to be critical; stereospecific alignment of the two sequences on the DNA helix is not essential, suggesting that they are independent sites.[27] CArG box type sequences are found in the proximal promoters of other muscle genes, and in some non-muscle genes where they also play a regulatory role, so that they are not activators of muscle specific transcription per se.

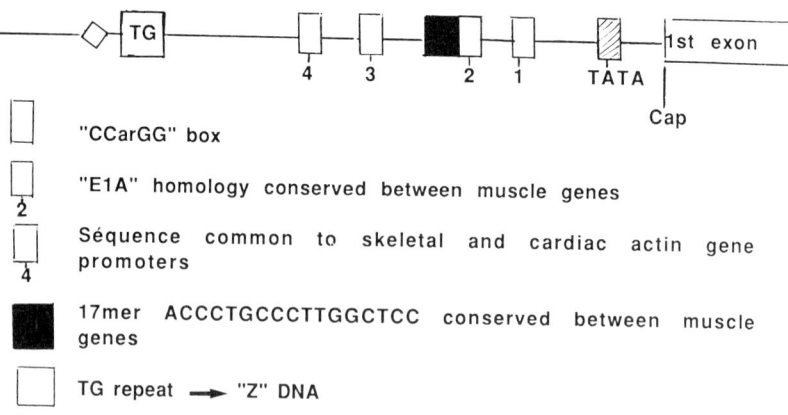

Fig. 5 A schematic representation of potential regulatory sequences in the mouse cardiac actin gene promoter.

Other more extensive conserved sequences have been pointed out between myosin alkali light chain and actin promoters expressed in fetal skeletal muscle,[3] and also between myosin alkali light chain genes expressed in striated muscle.[29] However, the functional significance of these common regions remains to be demonstrated. A special category of positive regulatory sequences are enhancers, which, as their name implies, greatly increase the basal level of expression of a gene. Characteristically, their activity is independent of their position and orientation in relation to the gene. There is some evidence that enhancer-like elements are present in the 5′ flanking sequences of a number of muscle genes, such as that for skeletal actin.[26] One of the best characterized examples is that of the creatine phosphokinase gene.[25] Here a muscle specific enhancer element has been identified which is situated just over a kilobase upstream of the transcription start site. The proximal 5′ flanking region of the gene promotes muscle specific activity on its own but this is increased about a hundred fold in the presence of the enhancer. With a heterologous promoter the creatine phosphokinase enhancer will confer muscle specificity. It has some homology with other well characterized enhancers such as that of the virus SV40, but is a distinct sequence, as indeed its muscle specificity suggests.

In a number of cases sequences have been identified within or around a muscle gene, outside the 5′ proximal region which confer muscle specificity and/or increase the level of expression of the gene in transfection assays. Two regulatory regions have been identified for the fast skeletal troponin I gene.[30] Appropriate developmental regulation, with transcriptional activation as muscle cells differentiate, requires a sequence present in the first intron of the gene, which acts in concert with the 5′ proximal region. This sequence has some of the properties of an enhancer. In the case of the myosin alkali light chain gene which encodes the two light chains of adult fast skeletal muscle (see Fig. 1) the two 5′ sequences proximal to the first coding exons of the two transcripts promote transcription in differentiated muscle cell cultures, but that for MLC1F in particular only does so at a low level.[21] In birds sequences further upstream appear to be important in boosting the level of expression.[31] In mammals, the striking finding is that there is a region in the 3′ flanking sequence of the gene, more than 20 kilobases downstream of the proximal promoter for the first transcript which greatly increases the muscle specific transcriptional activity of this promoter.[32] The 3′ region has many of the characteristics of a classical enhancer element; its functioning for example is independent of its position. The region involved is complex and appears to be composed of a number of sequence elements. The troponin I, creatine phosphokinase and myosin light chain sequences represent distinct muscle specific enhancers. Other regulatory elements situated outside the 5' proximal promoter sequence are undoubtedly involved in the expression of many muscle genes. The 'leakiness' seen with some of these regions on transfection in the tissue culture systems indicates this, and even where regulation is apparently muscle specific the level of expression attained may not be equivalent to that required in vivo. One way in which to obtain an indication of sequences implicated in muscle gene expression in tissue culture systems or in vivo is to use hypersensitivity to the enzyme DNase I as a criterion. Thus, for example, a number of DNase I hypersensitive sites of potential regulatory significance have been described around the γ subunit of the mouse acetylcholine receptor gene, two of which are located at a considerable distance 5' and 3' of the gene.[33]

At present very few experiments have been carried out to look directly at muscle gene regulation in vivo. It has however been shown by introducing genes for the myosin regulatory light chain and skeletal actin from the rat into transgenic mice that muscle specific expression can be obtained and moreover that, in the case of skeletal actin, this is observed with the 5' proximal promoter region alone, attached to an indicator gene.[34] Experiments of this kind in Xenopus have shown that the 5' proximal region of the cardiac actin gene is sufficient for expression in developing muscle.[35] Such experiments give an estimation of the sequence requirements for tissue specific expression in vivo, not only in the adult, but also, potentially, during muscle formation and maturation in the transgenic animal. It is less apparent how to address the question of quantative levels of expression in vivo, when the endogenous gene is also being expressed. In the longer term, the use of homologous recombination[5] in mammals to eliminate endogenous gene activity, should permit testing of a heterologous gene construction for full biological regulation, both qualitative and quantitative.

Transcriptional factors that regulate muscle gene expression

The sequences which have been shown to be functional in promoting muscle gene expression are present in or around the gene concerned and therefore by definition act in '*cis*'. Such DNA sequences interact with transcriptional factors present in the nucleus which act in '*trans*'. No *bona fide* transcriptional factor has yet been purified from muscle, but in a number of cases now it has been demonstrated that nuclear proteins are binding to regulatory sequences. Indeed such binding is in itself strongly suggestive of a regulatory function and complements the information obtained by transfection experiments. The two techniques most frequently used to study binding are gel retardation and DNA foot printing. In the former the migration of a labelled DNA fragment is examined on a non denaturing gel in the presence and absence of nuclear extract; protein binding to the fragment slows its migration in the gel. In the latter, protein binding to a given sequence element interferes with the chemical modifications of DNA sequence reactions to produce a characteristic gap in the sequence gel where the nucleotides binding the protein are located (Fig. 6).

Protein binding has been demonstrated for the first CArG box of the human cardiac actin gene, and the nucleotides involved in and around the CArG motif have been defined[28,36]. The second CArG box region of this gene[37] and also the CArG elements identified in the human skeletal actin gene[38] compete for binding with the first CArG sequence of the cardiac actin gene, suggesting that a common transcriptional factor is involved. Binding activity is seen with nuclear extracts from muscle cells, but also from non-muscle cells,[37,38] raising the question of the specificity of this phenomenon. Indeed, as already mentioned, similar CArG sequences are present in non-muscle genes and it has been shown, for example, that the serum response element (SRE) in the c-*fos* gene competes for binding activity with the cardiac actin CArG sequence, in extracts from non-muscle and muscle cells. Furthermore the complexes co-migrate on retardation gels suggesting that they bind the same factor[28,36] (Fig. 6). A number of possibilities can be evoked to try to explain why the proximal promoter of the cardiac actin gene is nevertheless expressed at a much higher level in differentiated muscle cells. There may be a muscle specific factor(s) which interacts with another region of this sequence, although the data currently available point to the importance of the CArG box. In muscle another factor may intervene to interact with the CArG binding factor (CBF), or this protein may undergo tissue specific secondary modifications. Alternatively the factors in muscle and non-muscle cells may be different although of similar size and with a common binding domain.

In this context, experiments on the proximal promoter of the chicken α-skeletal actin gene are important. A CArG box-like sequence is present at about 78 nucleotides upstream from the

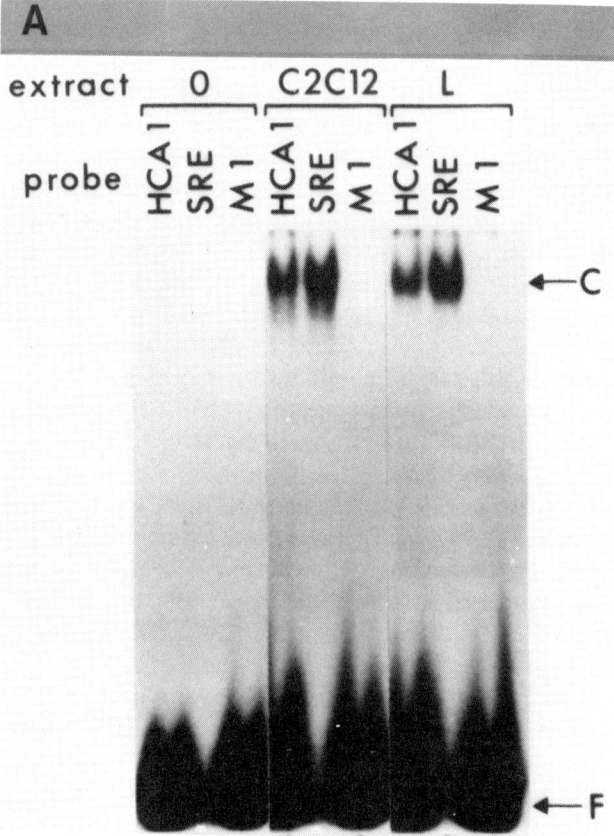

Fig. 6 Examples of a gel retardation experiment (**A**) and a footprint (**B**) on the human cardiac actin gene promoter[28]

A. Extracts from muscle cells ($C_2 C_{12}$) or non-muscle cells (L) were incubated with labelled oligonucleotides containing the first CCArG box (HCA1) or a mutated version of this AAArTT (M_1) from the cardiac actin gene, or the CCArGG box from the serum response element (SRE) found upstream of the c-*fos* gene. Complexed DNA (C) indicates factor binding. F indicates free (unbound) DNA **B**. Footprinting by methylation interference indicates where proteins are binding to the G residues of each strand (labelled top or bottom) of the CCArGG box sequence. Lanes F and B represent free DNA (control) and bound DNA (experiment) from the retardation gel with muscle cell extracts. Lane G represents marker degradation products of the probe sequence cleaved at guanine (G) residues. G residues missing or decreased in the bound (B) samples are shown by closed circles, and indicate protein binding in this position.

transcription initiation site, which binds to a factor in extracts from muscle and a number of different non-muscle cells. However in this case the major complex formed with muscle cell extracts migrates differently on retardation gels, although the DNA footprint is the same in all cases.[39] This experiment would suggest that different proteins or protein complexes are present in muscle

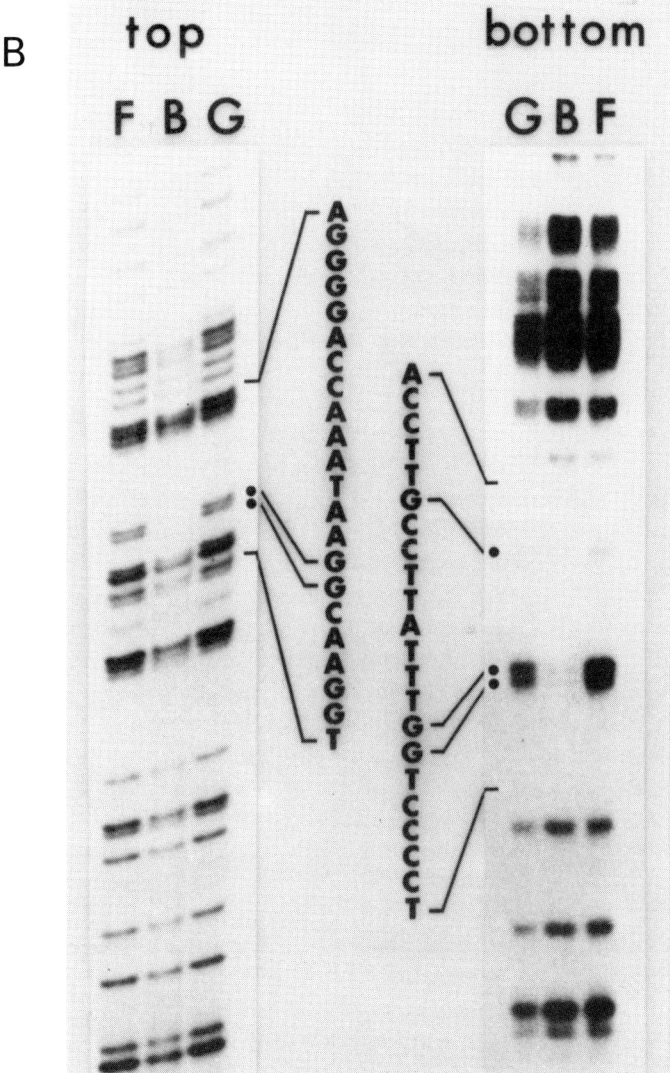

Fig. 6B (see opposite)

compared to non muscle cells, and that these are capable of binding to the same sequences. Both undifferentiated myoblasts and differentiated myotubes have the 'muscle-type' binding activity, so clearly this alone cannot explain the activation of the proximal promoter when differentiated myotubes form. It is possible that the chromatin configuration around the actin gene is such that the site is not accessible for binding in myoblasts, or indeed in non-muscle cells, and that the 'non-muscle' type binding activity only occurs in vitro.

A cardiac myosin regulatory light chain provides an example of a muscle gene where protein binding has been demonstrated to regulatory elements which are not CArG sequences. In this case binding activity was also found in non-muscle cell extracts, but at several fold lower concentration than that in heart extracts. A component is also present in non-muscle cell extracts which binds to a region shown to act as a negative control element, suppressing the activity of this gene in non-muscle cells.[22]

Experiments are in progress in several laboratories at present to look at nuclear protein binding to regulatory elements for a number of different muscle genes. In the next year or so, the questions of whether muscle specific factors are involved, and of whether some of these factors are common to several muscle genes should become much clearer. Hopefully it will prove possible to develop in vitro transcriptional systems for muscle, so that what is initially defined as a DNA binding activity can be tested for transcriptional function. Transcriptional factors for muscle genes will undoubtedly be isolated, and their genes in turn will become accessible for analysis. There are however, already, a number of factors, or candidate factors which have been characterized independently of their capacity to bind to muscle gene sequences. The most evident of these is thyroid hormone, which acts via the family of thyroid hormone receptor proteins. It is not clear whether thyroid hormone has a direct effect on the expression of many muscle genes, although it is important during muscle development. However in cardiac muscle it has been clearly demonstrated that DNA sequences required for thyroid hormone induction are present in the 5' flanking sequence of the α-cardiac myosin heavy chain gene.[40] Thyroid hormone receptor sequences are being cloned from muscle tissues at present. It has been shown that one such nuclear sequence is a thyroid hormone dependent transcriptional factor which binds to the responsive element of the α-myosin heavy chain gene to activate transcription.[41] Alternative splicing of thyroid hormone receptor gene transcripts produces factors with differing tissue specificities and transcriptional activities (e.g. Ref. 41) thus providing an example of how similar but distinct transcriptional factors may be generated.

Another class of sequences which has been isolated recently from rodent muscle cells in culture, is typified by Myo D1 (Myogenic Determination factor 1)[42] which is capable on transfection of converting a parental precursor cell type (the C3H 10T1/2 mouse cell line) to the myogenic lineage, and which is normally expressed in both myoblasts and myotubes, but not in the parental precursor cell type. Myo D1 encodes a protein which has sequence motifs—a C-*myc* domain and a zinc finger domain—characteristic of nuclear DNA binding proteins. A second sequence of this kind, myogenin[43] also converts 10T1/2 cells to the myogenic lineage and has similar sequence motifs to Myo D1, although it is clearly a different protein, with a rather different pattern of expression during myogenesis. Another sequence, myd, isolated as a genomic clone, also functions as a myogenic differentiation factor.[44] Further factors have been isolated from human muscle on the basis of their homology with Myo D1.[45] It is not yet clear how these factors act, but the possibility that they may interact directly with regulatory elements in muscle genes is an attractive hypothesis. Preliminary experiments with Myo D1 suggest that binding to some muscle regulatory sequences may take place. These factors may therefore correspond to a class of muscle specific transcription factors and their isolation as DNA clones if this is the case circumvents the tedious and difficult steps of protein purification required to obtain such factors.

CONCLUSIONS

In this brief review of the control of muscle gene expression, splicing mechanisms have been discussed, and current work on transcriptional regulation has been reviewed in some detail. It is clear from transcriptional run on experiments with isolated nuclei from cultured muscle cells that the myoblast to myotube transition is accompanied by the transcriptional activation of many muscle genes[46] and attention has tended to focus on this step. This reflects the accessibility of the tissue culture system and the very considerable recent progress in technical approaches to the study of transcription. However, the question of how developmental changes in muscle gene expression in vivo are regulated remains largely open. There are indications, even from the tissue culture studies, that post-transcriptional regulation operates in older myotubes to determine the level of muscle mRNAs, probably at the level of mRNA stability.[46] There are also indications from in vivo studies on mice carrying a mutant cardiac actin gene locus where a reduction in actin mRNA levels by as much as a factor of five does not result in reduced actin protein, that mechanisms affecting translation and/or protein stability may also be very important in maintaining a muscle phenotype.[3] These aspects of muscle gene regulation will not be discussed here, but they undoubtedly merit more attention. In the immediate future, rapid progress will probably continue to be made in the analysis of the production and processing of the primary transcript, in terms of the factors regulating these phenomena. In the longer term understanding how proteins interact with a sequence and how this regulates transcriptional activity (*see* Ref. 47 for a discussion of general mechanisms) becomes a question for physical chemists rather than biologists. How gene regulatory factors are modulated under different physical conditions, and how their genes are regulated remains the domain of the molecular biologist. This raises the quandary of genes regulating genes *ad infinitum*! An indication of how this may be resolved is perhaps given by recent studies on the genes encoding AP1 and c-*fos* which act as general transcriptional factors, and appear to function as auto regulators of their own promoters in conjunction with other factors (e.g. *see* Ref. 48). It is also becoming clear that secondary modifications of transcriptional factors are likely to be important in modulating their activity (*see* Ref. 49).

ACKNOWLEDGEMENTS

M Buckingham's Laboratory is supported by grants from the Pasteur Institute, the Centre National de la Recherche Scientifique, the Institut National de la Santé et de la Recherche Médicale, the Association pour la Recherche sur le Cancer, the Commission des Communautés Européennes, the North Atlantic Treaty Organisation, and the Muscular Dystrophy Associations of America. M B is grateful to Paul Barton and Benoît Robert for critical comments on the manuscript and to Marc Fiszman for discussion of splicing. Françoise Tuy very kindly provided the material used in Figure 6.

REFERENCES

1 Buckingham ME. Actin and myosin multigene families: their expression during the formation of skeletal muscle. In: Essays in Biochemistry 1985; 20: 77–109

2 Vandekerckhove J, Weber K. The complete amino acid sequence of actins from bovine aorta, bovine heart, bovine fast skeletal muscle, and rabbit slow skeletal muscle. Differentiation 1979; 14: 123–133

3 Garner I, Minty AJ, Alonso S, Barton PJ, Buckingham ME. A 5' duplication of the α-cardiac actin gene in BALB/c mice is associated with abnormal levels of α-skeletal actin mRNAs in adult cardiac tissue. EMBO J 1986; 5; 10: 2559–2567

4 Mahaffey JW, Coutu MD, Fyrberg EA, Inwood W. The flightless Drosophila mutant *raised* has two distinct genetic lesions affecting accumulation of myofibrillar proteins in flight muscles. Cell 1985; 40: 101–110

5 Mansour SL, Thomas KR, Capecchi MR. Disruption of the proto-oncogene *int-2* in mouse embryo-derived stem cells: a general strategy for targeting mutations to non-selectable genes. Nature 1988; 336: 348–352

6 Jacob F. Evolution and tinkering. Science 1977; 196(4295): 1161–1167

7 Breitbart RE, Andreadis A, Nadal-Ginard B. Alternative splicing: a ubiquitous mechanism for the generation of multiple protein isoforms from single genes. Ann Rev Biochem 1987; 56: 467–495

8 Robert B, Daubas P, Akimenko M-A, et al. A single locus in the mouse encodes both myosin light chains 1 and 3, a second locus corresponds to a related pseudogene. Cell 1984; 39: 129–140

9 Helfman DM, Cheley S, Kuismanen E, Finn LA, Yamamaki-Kataoka Y. Nonmuscle and muscle tropomyosin isoforms are expressed from a single gene by alternative RNA splicing and polyadenylation. Mol Cell Biol 1986; 6: 3582–3595

10 Reinach F, MacLeod A. Tissue-specific expression of the human tropomyosin gene involved in the generation of the *trk* oncogene. Nature 1986; 322: 648–650

11 Wieczorek DF, Smith CWJ, Nadal-Ginard B. The rat α-tropomyosin gene generates a minimum of six different mRNAs coding for striated, smooth, and nonmuscle isoforms by alternative splicing. Mol Cell Biol 1988; 8(2); 679–694

12 Barton PJR, Buckingham ME. The myosin alkali light chain proteins and their genes. Biochem J 1985; 231: 249–261

12a Barton PJR, Robert B, Cohen A, et al. Structure and sequence of the myosin alkali light chain gene expressed in adult cardiac atria and fetal striated muscle. J Biol Chem 1988; 263(25); 12669–12676

13 Breitbart RE, Nadal-Ginard B. Developmentally induced, muscle-specific *trans* factors control the differential splicing of alternative and constitutive troponin T exons. Cell 1987; 49: 793–803

14 Gahlmann R, Troutt AB, Wade RP, Gunning P, Kedes L. Alternative splicing generates variants in important functional domains of human slow skeletal troponin T. J Biol Chem 1987; 262(33): 16122–16126

15 Cooper TA, Cardone MH, Ordahl P. Cis requirements for alternative splicing of the cardiac troponin T pre-mRNA. Nucleic Acids Res 1988; 16: 8443

16 Dickson G, Hilary JG, Barton CH, et al.. Human muscle neural cell adhesion molecule (N-CAM): identification of a muscle-specific sequence in the extracellular domain. Cell 1987; 50: 1119–1130

17 Hernandez N, Keller W. Splicing of in vitro synthesized messenger RNA precursors in HeLa cell extracts. Cell 1983; 35: 89–99

18 Robert B, Barton P, Minty A, et al. Investigation of genetic linkage between myosin and actin genes using an interspecific mouse back-cross. Nature 1985; 314; 6007: 181–183

19 Heidmann O, Buonanno A, Geoffroy B et al. Chromosomal localization of muscle nicotinic acetylcholine receptor genes in the mouse. Science 1986; 235: 866–868

20 Weydert A, Daubas P, Lazaridis I, et al. Genes for skeletal muscle myosin heavy chains are clustered and are not located on the same mouse chromosome as a cardiac myosin heavy chain gene. Proc Natl Acad Sci USA 1985; 82: 7183–7187

21 Daubas P, Klarsfeld A, Garner I, Pinset C, Cox R, Buckingham ME. Functional activity of the two promoters of the myosin alkali light chain gene in primary muscle cell cultures: comparison with other muscle gene promoters and other culture systems. Nucleic Acids Res 1988; 16(4): 1251–1271

22 Braun T, Tannich E, Buschhausen-Denker G, Arnold H-H. Promoter upstream elements of the chicken cardiac myosin light chain 2-A gene interact with trans-acting regulatory factors for muscle specific transcription. Mol Cell Biol 1989 (In press)

23 Mar JH, Antin PB, Cooper TA, Ordahl C. Analysis of the upstream regions governing expression of the chicken cardiac troponin T gene in embryonic cardiac and skeletal muscle cells. J Cell Biol 1988; 107: 573–585

24 Bouvagnet PF, Strehler EE, White GE, Strehler-Page M-A, Nadal-Ginard B, Mahdavi V. Multiple positive and negative 5' regulatory elements control the cell-type-specific expression of the embryonic skeletal myosin heavy-chain gene. Mol Cell Biol 1987; 7: 4377–4389

25 Jaynes JB, Johnson JE, Buskin JN, Gartside CL, Hauschka SD. The muscle creatine kinase gene is regulated by multiple upstream elements, including a muscle specific enhancer. Mol Cell Biol 1988; 8: 62–70

26 Muscat GEO, Kedes L. Multiple 5'-flanking regions of the human α-skeletal actin gene synergistically modulate muscle-specific expression. Mol Cell Biol 1987; 7(11): 4089–4099

27 Miwa T, Kedes L. Duplicated CArG box domains have positive and mutually dependent regulatory roles in expression of the human α-cardiac actin gene. Mol Cell Biol 1987; 7(8): 2803–2813

28 Phan-Dinh-Tuy F, Tuil D, Schweighoffer F, Pinset C, Kahn A, Minty A. The 'CC.Ar.GG' box. Eur J Biochem 1988; 173: 507–515

29 Cohen A, Barton PJR, Robert, B, Garner I, Alonso S, Buckingham ME. Promoter analysis of myosin alkali light chain genes expressed in mouse striated muscle. Nucleic Acids Res 1988; 16(21): 10037–10052

30 Konieczny SF, Emerson C. Jr. Complex regulation of the muscle-specific contractile protein (Troponin I) gene. Mol Cell Biol 1987; 7; 9: 3065–3075

31 Shirakata M, Nabeshima Y-I, Konishi K, Fujii-Kuriyama Y. Upstream regulatory region for inducible expression of the chicken skeletal myosin alkali light-chain gene. Mol Cell Biol 1988; 8; 6: 2581–2588

32 Donoghue M, Ernst H, Nadal-Ginard N, Rosenthal N. A muscle-specific enhancer at the 3' end of the myosin light chain 1/3 gene locus. Genes Dev 1988; 2: 1779–1790

33 Crowder CM, Merlie JP. DNase I-hypersensitive sites surround the mouse acetylcholine receptor γ-subunit gene. Proc Natl Acad Sci USA 1986; 83: 8405–8409

34 Shani M. Tissue-specific and developmentally regulated expression of a chimeric actin-globin gene in transgenic mice. Mol Cell Biol 1986; 6; 7: 2624–2631

35 Mohun TJ, Garrett N, Gurdon JB. Upstream sequences required for tissue-specific activation of the cardiac actin gene in Xenopus laevis embryos. EMBO J 1986; 5; 12: 3185–3193

36 Gustafson TA, Miwa T, Boxer LM, Kedes L. Interaction of nuclear proteins with muscle-specific regulatory sequences of the human cardiac α-actin promoter. Mol Cell Biol 1988; 8(11): 4110–4119

37 Miwa T, Boxer LM, Kedes L. CArG boxes in the human cardiac α-actin gene are core binding sites for positive trans-acting regulatory factors. Proc Natl Acad Sci USA 1987; 84: 6702–6706

38 Muscat GEO, Gustafson TA, Kedes L. A common factor regulates skeletal and cardiac α-actin gene transcription in muscle. Mol Cell Biol 1988; 8: 4120–4133

39 Walsh K, Schimmel P. DNA-binding site for two skeletal actin promoter factors is important for expression in muscle cells. Mol Cell Biol 1988; 8(4): 1800–1802

40 Gustafson TA, Markham BE, Bahl JJ, Morkin E. Thyroid hormone regulates expression of a transfected α-myosin heavy-chain fusion gene in fetal heart cells. Proc Natl Acad Sci USA 1987; 84: 3122–3126

41 Izumo, Mahdavi V. Thyroid hormone receptor α-isoforms generated by alternative splicing differentially activate myosin HC gene transcription. Nature 1988; 334: 539–541

42 Davis RL, Weintraub H, Lassar AB. Expression of a single transfected cDNA converts fibroblasts to myoblasts. Cell 1987; 51: 987–1000

43 Wright WE, Sassoon DA, Lin VK. Myogenin, a factor regulating myogenesis, has a domain homologous to Myo D1. Cell 1989; 56: 607–617

44 Pinney DF, Pearson-White S, Konieczny SF, Latham KE, Emerson CP Jr. Myogenic lineage determination and differentiation: evidence for a regulatory gene pathway. Cell 1988; 53: 781–793

45 Braun T, Buschhausen-Denker G, Bober E, Arnold HH. A novel human muscle factor related to but distinct from MyoD1. EMBO J 1989 (in press)

46 Cox R, Buckingham ME. 1989 (manuscript in preparation)

47 Ptashne M. How eukaryotic transcriptional activators work. Nature 1988; 335: 683–689

48 Sassone-Corsi P, Sisson JC, Verma IM. Transcriptional autoregulation of the proto-oncogene fos. Nature 1988; 334: 314–318

49 Jackson SP, Tjian R. O-glycosylation of eukaryotic transcription factors: implications for mechanisms of transcriptional regulation. Cell 1988; 55: 125–133

British Medical Bulletin (1989) Vol. 45, No. 3, pp. 630–643

Duchenne/Becker muscular dystrophy:

A SHORT OVERVIEW OF THE GENE, THE PROTEIN, AND CURRENT DIAGNOSTICS

L M Kunkel
E P Hoffman
Pediatrics, Harvard Medical School. Howard Hughes Medical Institute. Children's Hospital, Boston, USA

The goal of this review is to provide a concise overview of the current state of knowledge of both the gene which, when defective, results in Duchenne muscular dystrophy (DMD), and the normal and abnormal protein products of this gene. The diagnostic tools, both prenatal and postnatal, which have resulted from this research will also be presented. More comprehensive discussions of many aspects of this review can be found elsewhere.[1,2]

REVERSE GENETICS AND THE CLONING OF THE DMD GENE

By the early 80's the position of the DMD gene had been established as Xp21. This was accomplished by both linkage of DMD mutations to Xp21 RFLP allele detecting probes[3] and the study of certain unusual patients.[4] Specifically, a few females had been described who exhibited many of the characteristic clinical features of DMD. Many of these females were shown by cytogenetic analysis to be balanced X;autosome translocation carriers, with each having a common translocation breakpoint in Xp21. In 1985 a male was described who exhibited three X-linked disease phenotypes including DMD.[5] Cytogenetic analysis of mitotic cells of this patient indicated that this patient had a deletion of Xp21 materials which was probably the basis by which he manifested the disorders. A gross deletion of the Xp21 region of his X chromosome was confirmed by the absence of the cloned DNA segment 754.[5] Presumably, any cloned segment which could be shown to be absent

0007–1420/89/0045–0630/$10.00

from DNA isolated from this patient might be within or nearby one of the loci being disrupted to yield the phenotypes observed in the patient. A competition reassociation and cloning strategy was used to specifically obtain DNA segments absent from the patient's DNA.[6] One way to use these cloned segments was to search the DNA of multiple DMD patients for a detectable alteration such as a deletion or rearrangement. Knowledge from other disorders indicated that approximately 5% of patients afflicted with any genetic disorder could be assumed to exhibit a mutation in the specific gene involved. Originally, 57 DNA samples isolated from unrelated patients with DMD were tested for the absence of one or more of eight cloned segments from Xp21.[7] One clone, pERT87, was found to be missing from DNA isolated from 5 of these patients, with no DNA samples isolated from normal individuals showing such a deletion. The ability of pERT87 to detect deletion mutations in a subset of patients made this particular clone a likely candidate to be nearby or within the DMD locus.

The small pERT87 clone was expanded by chromosome walking in human recombinant phage libraries.[8,9] The intention was to define the extent of the deletions and possibly define regions important towards the normal expression of the DMD gene product. Ultimately, a 220 kb contiguous block of cloned DNA was obtained within which many deletions exhibited breakpoints. Indeed, the deletions were found to extend in different directions relative to the block of cloned DNA; strong evidence that there might be determinants of the gene within. During the course of chromosome walking, multiple northern blots were prepared and tested by hybridization with subclones from the region. No evidence of specific transcripts could be obtained so an alternative approach was taken. The coding regions of most loci tend to be conserved at the nucleotide sequence level whereas non-coding regions are generally not conserved. A systematic search for nucleotide sequence conservation was undertaken and two regions were found to cross-hybridize to all mammalian DNA samples tested.[9] The DNA of both regions was sequenced and the sequences searched for the consensus nucleotides for RNA processing. Two small segments were found which might be potential exons of a gene. Upon testing for expression as RNA, one conserved segment detected a very large low abundance RNA in fetal skeletal muscle. Poly A + RNA was prepared and used to construct a cDNA library which was screened with both conserved regions. Positively hybridizing cDNAs were plaque purified and DNA prepared. The largest insert of approximately 1 kb hybridized to multiple restriction fragments in total DNA which each could be mapped to Xp21. Hybridization back to walked phage DNA samples of the 220 kb contiguous block identified eight potential exons. With only 1/16th of the large transcript cloned as cDNA, the exons for this portion of the gene were spaced over 130 kb of genomic DNA. If the remainder of the transcript was encoded by exons spread over similar distances, then the gene was postulated to be nearly 1 million base pairs in size[9].

NORMAL GENE STRUCTURE

The DMD gene is by far the largest genetic locus characterized to date. The gene contains at least 65 exons distributed over 2 500 000 base pairs of the human X chromosome.[1,10,11] The gene can be considered to have been very highly conserved through evolution, not only with regards to direct sequence homology, but also with regards to the complexity of the gene (number of exons) and size of the genomic locus.[12-14] The enormous size of the gene is probably at least partly responsible for the unusually high mutation rate of the gene, and thus the high frequency of DMD in all human populations,[15] and in mice.[16] The normal DMD gene is transcribed and spliced into a mRNA of approximately 14 000 bp, which has been found as a low abundance component (0.01–0.001%) of all muscle tissue poly-A + RNA, and in much lower amounts in many non-muscle tissues.[9,12,17-20] It is not clear, at this point in time, why the DMD gene has maintained such a large size, with over 2 million base pairs of non-translated DNA (introns). As some of the individual introns are thought to be over 200 000 bp in length, it is tempting to speculate that smaller genes may reside within some of the DMD introns. There is, however, no evidence to date suggesting that additional transcription units exist within the DMD gene.[21]

ABNORMAL DMD GENE STRUCTURE

The high mutation rate of the DMD gene ensures that most mutations of the gene are independently derived. Thus, the very large patient population afflicted with DMD would be expected to exhibit a plethora of unrelated, potentially 'unique' genetic mutations, each of which partially or completely disables the gene. At least 60% of affected patients have been found to possess deletions of one or more exons of the DMD gene, and indeed these deletions have been found to be very heterogeneous both with regard to the extent of sequences deleted, and the position of the deletion within the gene.[10] There are a few apparent 'hot spots' for the initiation of deletion, with three of the more than 60 introns responsible for initiating over 40% of all deletion mutations.[10] It is not clear at this point whether there are specific sequences within these introns that are highly prone to deletion initiation, or whether these introns are extremely large, and thus simply a large target for mutation. The complexity and size of the DMD gene has made it technically difficult to investigate the mechanism of genetic mutation in the 35% of patients not exhibiting a deletion. Specifically, the possible existence of point mutations in the protein coding region or in splice site acceptor or donor sequences remains to be investigated. There have, however, been a number of duplication mutations reported.[8,22]

The isolation of the complete DMD gene cDNA (14 kb) permits the relatively simple visualization of the 65 unique exons which together define the coding sequences of the 2.5 mega-bp gene. Thus, afflicted patients can be easily scored for the presence or absence of each of the 65 exons. In this manner, the precise mutation can be defined in the 65% of patients exhibiting a deletion of one or more exons. Using the cDNA, the specific mutations have been defined in hundreds of patients. As different mutations of the DMD/BMD gene can result in a wide range of clinical severities, the correlation of deletion position and extent within the gene with the clinical severity of individual patients held promise regarding the structure and function of the protein product of the gene. It quickly became clear, however, that there was no such clear correlation between the position/extent of deletions and the severity of the clinical phenotype. An 'in-frame/out-of-frame' hypothesis was forwarded as an alternative explanation of how certain deletions could result in a severe clinical phenotype, while other overlapping and often larger deletions could result in a mild clinical phenotype.[23] This hypothesis involved the ability of the deleted gene to produce a functional 'translatable' mRNA despite the mutation: 'DMD deletions' resulted in a 'nonsense' type of mutation when the remaining exons were brought together by mRNA splicing, while 'BMD-deletions' resulted in mRNA molecules which were capable of directing the translation of a semi-functional protein.[2,23] Thorough and refined molecular analysis of specific deletions has substantiated the validity of this bi-modal hypothesis.[23,24] This hypothesis has been further substantiated at the protein level.[25] There are, however, some apparent exceptions to this rule.[26]

THE PROTEIN PRODUCT: DYSTROPHIN

The primary biochemical defect (abnormal or missing protein) responsible for DMD/BMD had eluded scientists for many decades. The identification of the DMD/BMD gene and the isolation of its cDNA coding sequences initiated two fruitful avenues of research into the nature of the biochemical defect; the theoretical prediction and analysis of the primary amino acid sequence encoded by the gene, and the production of antibodies directed against the hypothetical protein. Antisera were produced by first directing bacteria to synthesize large fragments of the predicted protein in large quantities and at high purity.[27] These large fragments were then used as antigens for immunization, and the resulting antibodies affinity purified with these same bacteria-produced antigens. A large (400 kd), low abundance (0.01% of cellular protein) protein species was recognized by all antisera in normal myogenic cells, which was named dystrophin.[27,28] This protein appears to be highly conserved through evolution, with the antisera detecting dystrophin homologues in mouse and chicken, among other species.[13,28] The gene size and the amino acid and DNA sequences also appear to be highly conserved, with dystrophin-like genes detectable down to amphibians using human cDNA sequences as molecular probes.[13]

Using various antibody preparations, dystrophin was found to be associated with membrane fractions from muscle, specifically with the t-tubules of myofibers.[29,30] Parallel immunofluorescent studies using these same antisera and other, independently derived antisera, localized dystrophin at the periphery of myofibers, indicating an association with the plasma membrane.[31–33] Though the cytological results have been interpreted as being contradictory to the previous biochemical data,[32] the difficulty of visualizing the t-tubules at the light microscopic level, particularly given the low abundance of dystrophin, and the contiguous nature of the plasma membrane and t-tubules, suggests that these results are instead complimentary, indicating that dystrophin is likely to adhere to both the plasma membrane and t-tubules. Indeed, immuno-electron microscopy then directly visualized dystrophin on the internal (cytoplasmic) face of the plasma membrane, and possibly the contiguous t-tubules, in a periodic manner suggesting some type of network arrangement.[34] Dystrophin is found in all types of normal, terminally differentiated myogenic cells, including vascular and visceral smooth muscle. Dystrophin is also thought to be specifically expressed in neurons.[28] There are probably a number of tissue-specific isoforms of dystrophin, as evidenced both by immunoblot analysis[28] and extensive PCR (polymerase-chain-reaction) analyses of dystrophin mRNA in various tissues.[35]

Amino acid sequence analysis of dystrophin has shown that the protein is comprised of approximately 3600 amino acids, arranged into four relatively distinct domains.[36] The first three of these domains appear to be highly related to analogous domains of non-muscle alpha-actinin, a cytoskeletal protein found in many cell types in many organisms.[37] The first of these domains is an amino-terminal 240 amino acid region which, in the case of cytoskeletal alpha-actinin, has been shown to bind filamentous actin.[38] The second domain is a series of putative triple alpha-helixes thought to adopt a rod type of tertiary structure.[36] This domain is comprised of approximately 400 amino acids (4 repeats) in alpha-actinin, and approximately 2700 amino acids (25 repeats) in dystrophin. The third region shared by both proteins is the cysteine-rich, carboxy-terminal domain of alpha-actinin which appears related to the penultimate domain of dystrophin. This cysteine-rich region in dystrophin seems to extend beyond the 140 amino acids of similar ity with alpha-actinin, and does not appear to contain the Ca^{2+} binding sites (EF hands) found in this region in alpha-actinin. The fourth, 420 amino acid carboxy-terminal domain of dystrophin does not appear related to any previously characterized protein.

The alpha-actinins are a biophysically defined group of proteins which include both the cytoskeletal (non-muscle) alpha-actinins, described above, and also the myofibrillar (muscle) alpha-actinins which are the major components of the Z-line in myofibrillar sarcomeres. Thus, while all alpha-actinins are structurally related and are believed to bind actin filaments, the different subcellular localizations of myofibrillar and cytoskeletal alpha-actinins suggests different specific functions. Despite the sequence similarities between cytoskeletal alpha-actinin and dystrophin described above, dystrophin has been shown to be immunologically related to myofibrillar (muscle) alpha-actinin and not cytoskeletal alpha-actinin.[39] In addition, dystrophin shares sequence similarities with spectrin, the major component of the erythroid membrane cytoskeleton. As many of the biophysical characteristics of the alpha-actinins are thought to be shared by spectrin, it is likely that dystrophin shares the biophysical characteristics of these proteins. Thus, it is hypothesized that dystrophin, like spectrin and the alpha-actinins, is an antiparallel dimer with an extended (125 nm) central rod domain, with more globular amino- and carboxy-terminal domains. Like spectrin and cytoskeletal (non-muscle) alpha-actinin, dystrophin is membrane associated despite the lack of any transmembrane domains. Thus, dystrophin is thought to interact with integral membrane proteins via its terminal domains, a super-structure which has been described for spectrin in the erythrocyte.[40]

Thus, the current state of knowledge concerning the cellular function of the DMD gene product, dystrophin, is that it is most likely a component of the membrane cytoskeleton of myogenic, and probably neuronal, cells. Fortunately, there is an extensive literature documenting plasma membrane defects in DMD myofibers.[41] It remains to be determined, however, whether dystrophin mediates this membrane damage directly, or indirectly via the integral membrane proteins it should normally interact with. In this context, it is interesting to note the paucity of inter-plasma membrane orthogonal arrays in both Duchenne dystrophy and mdx mouse muscle.[42,43] It is quite possible that these orthogonal arrays represent the proteins to which dystrophin is normally bound.

ANIMAL MODELS OF DMD AND COMPARATIVE PATHOPHYSIOLOGY

Severe DMD has been shown to be the result of the complete, or nearly complete, absence of dystrophin.[24,44] In addition, dystrophin deficiency has been shown to be completely disease specific, such that no unrelated neuromuscular disease has been observed to exhibit reduced cellular abundance of dystrophin. Given the evolutionary conservation of the DMD gene and dystrophin, the conservation of X-linked genes among placental mammals, and the disease specificity of dystrophin deficiency observed in humans, it can be assumed that any X-linked animal dystrophy exhibiting dystrophin deficiency represents an animal model for human DMD. To date, dystrophin deficient mice, dogs and cats, all of which have an X-linked pattern of inheritance, have been described.[45–47] On the other hand, both the dystrophic chicken[27,28] and the dy mouse (Hoffman EP, Kunkel LM, unpublished results) have been shown to contain apparently normal levels of dystrophin, and are presumed non-homologous to human DMD. The homologous animal models (mdx mice, CXMD dogs, and the dystrophin-deficient cats) appear to be very instructive with regards to the pathophysiology of human DMD, both due to the similarities and dissimilarities between each of these models and DMD. The comparisons are most easily drawn at three different levels: biochemical comparisons, histological comparisons, and clinical (phenotypic) comparisons. At the biochemical level DMD humans and all three animal models can be considered to be identical: all exhibit the apparent complete absence of dystrophin. Comparison at the histological level is more difficult because there is considerable variability of the observed muscle histopathology depending on the age and the specific muscle group of the dystrophin-deficient organism studied. The histopathological features which are considered relatively invariant or 'static' in human DMD are also found in all animal models. These are focal fiber degeneration and regeneration, a large amount of muscle fiber splitting, increased proportions of centrally nucleated fibers, and considerable fiber size variation with both very large and very small fibers present in most sections.[46,48,49] An important histological feature which has been shown to be 'dynamic' in human DMD is the extent of connective tissue proliferation between muscle fibers (endomysial fibrosis). There is little obvious fibrosis in very young DMD patients, but with an increasing age fibrosis becomes predominant in many muscle groups, such that the majority of muscle can be replaced by fibrotic tissue in older DMD patients.[50] Such progressive fibrosis is also obvious in the dystrophin deficient dog. However, little or no fibrosis is evident in most muscles of the mouse and cat models. Concommitent with the fibrosis in humans and dogs is the gradual loss of muscle fibers, thought to be due to the inability of the muscle fibers to sustain adequate regeneration. The mouse and cat models, on the other hand, appear not to experience similar myofiber loss.

The clinical phenotype of human DMD and the three animal models seems to parallel the amount of myofiber loss (and fibrosis). For example, both young (less than 5 year) DMD-affected humans, young (less than 3 month) dystrophin deficient dogs exhibit very little fibrosis, very little myofiber loss, and very little detectable clinical weakness. With advancing age, fiber loss, fibrotic replacement of muscle, and clinical weakness all become apparent and dynamically progressive, leading to an early death in both human DMD and the dog model. The dystrophin deficient mice and cats, on the other hand, do not experience fiber loss, fibrosis or clinical weakness, and live a seemingly normal life. (It should be noted, however, that the dystrophin deficient cats studied exhibited 'stiffness' and were euthanized at a few years of age).

Thus, the comparison of human DMD with the animal models implies that the human pathophysiology can be separated into two 'phases', to 0 to 5 year static myopathic phase accompanied by little clinical weakness, and the 5 to 25 year dynamic progressive phase where concommitent myofiber loss and fibrosis result in clinical weakness and ultimately death. While the affected dogs progress through both phases of the disease in a manner strikingly similar to that of DMD-affected humans, dystrophin deficient mice and cats seem to remain largely within the static myopathic phase of the disease and thus exhibit little, if any, clinical weakness.

Complicating a full understanding of the pathophysiology of human DMD and the animal models is the recent implication of primary defects of the vasculature (vascular smooth muscle) and neurons.[27,28] Though many non-muscle cellular and physiological defects have been hypothesized for many years, the progressive nature of DMD has made it difficult to determine which of these defects are primary causes of the disease and which are secondary consequences. The validation of dystrophin deficiency in vascular smooth muscle and neurons suggests that primary defects of these tissues could be superimposed on primary defects of the dystrophin-deficient skeletal myofibers.

Though the dystrophin deficient mouse has been described as a poor model of DMD due to its lack of clinical weakness, instead it might represent an ideal experimental answer to a problem in studying human DMD which has been recognized for some time. The problem is best expressed in the conclusions of a comprehensive study of DMD histopathology published 20 years ago: 'These studies suggest that the presence of several distinct pathological features in muscle specimens from children with Duchenne muscular dystrophy may be caused by different reactive pathological processes which may obscure the specific inherited defect of this disease and aggravate muscle fibre destruction'.[50] Indeed, promising potential avenues towards therapy of Duchenne dystrophy have begun to be developed using the mdx mouse.[51]

CURRENT DIAGNOSTICS

Use of the complete DMD cDNA on genomic Southern analysis allows the detection of more than 65 unique exons distributed over the 2.5 million base pair DMD gene locus.[10] Since over 65% of DMD and BMD patients exhibit deletion mutations of one or more of these exons, the specific genetic lesion responsible for the disorder is thus easily detected in the majority of patients.[9,52,53] These observations have dramatically simplified prenatal diagnosis in most families. If an affected family member is shown to exhibit a deletion mutation, the fetus at risk in the family is simply tested for the presence or absence of that deletion. Not only does such analysis preclude the expensive and time consuming 'family linkage studies', the accuracy of such an approach nears 100%. Those families at risk which do not segregate a deletion must still be analyzed via more traditional linkage studies. The recent availability of intragenic RFLP detecting probes, however, has also increased the accuracy of this type of analysis.

Characterization of dystrophin in patients, though available for only a very short time, has already seen many varied practical applications, and promises to be a useful clinical tool of the neurologist. Visualization of dystrophin in individual patients is performed on small fragments of diagnostic muscle biopsies, by either immunofluorescent (cytological)[31,33,54,55] or immunoblotting (biochemical)[25,44] techniques. Both techniques have verified dystrophin deficiency in severe DMD in a large number of patients. Furthermore, the disease-specificity of dystrophin deficiency has been well established in these same reports, in that no patient with a neuromuscular disease unrelated to DMD/BMD exhibits abnormalities of dystrophin. Dystrophin testing has proven very valuable in the diagnosis of preclinical patients; young (0–4 year old) boys who have been noted to have grossly elevated serum creatine kinase levels, yet have no family history for a myopathy, and no clinical symptoms. Such young patients exhibit only a general myopathic process in their muscle, and without the typical clinical progression of DMD yet evident, it is difficult to reach an unambiguous diagnosis. Analysis of dystrophin content in these patients permits an early, and probably completely accurate diagnosis.[25,56,57] Though prenatal diagnosis is usually accomplished via the above described genetic testing, dystrophin testing has proven valuable in certain situations in combination with the genetic testing. Specifically, dystrophin testing can be performed on abortuses to confirm the genetic prenatal diagnosis, and also can be used to determine the X-chromosome bearing the disease gene in certain 'difficult' families.[58] Finally, successful detection of carrier females has been accomplished using dystrophin immunofluorescence,[54] though the intrinsic accuracy of such carrier detection remains to be determined.[59]

Finally, the heterogeneous and relatively unpredictable clinical progression of DMD, outliers, and BMD patients has made clinical investigations into the efficacy of therapeutic treatments problematic. Since it is difficult to predict the course of the clinical progression of untreated 'control' patients, it is likewise difficult to assess the beneficial or detrimental effects of any drugs on the experimental group. This situation has resulted in stringent statistical requirements for clinical trials in DMD.[60] With dystrophin testing, however, it will be possible to construct control and experimental groups exhibiting identical dystrophin abnormalities, all of whom would be expected to manifest very similar clinical progressions. Dystrophin testing of all patients included in a clinical study should increase the accuracy of the interpretations of such studies, and reduce the required stringency of previous statistical requirements.

In conclusion, much has been learned about the genetics, biochemistry, and pathophysiology of Duchenne and Becker muscular dystrophy within an extraordinarily short period of recent history. This progress has been the result of very fruitful collaborations of clinicians, patients, basic scientists, and funding institutions (especially Muscular Dystrophy Associations) throughout the world. Hopefully, this collaboration and consequential progress will continue, bringing the cure to the diseases within reach.

REFERENCES

1 Monaco AP, Kunkel LM. Cloning of the Duchenne/Becker muscular dystrophy locus. In: Harris H, Hirschorn K, eds. Advances in Human Genetics, vol.17. New York: Plenum, 1988: pp. 61–98
2 Hoffman EP, Kunkel LM. Dystrophin abnormalities in Duchenne/Becker muscular dystrophy. Neuron 1989; 2: 1019–1029
3 Davies KE, Pearson PL, Harper PS et al. Linkage analysis of two cloned DNA sequences flanking the Duchenne muscular dystrophy locus on the short arm of the human X chromosome. Nucleic Acids Res 1983; 11: 2303–2312
4 Boyd Y, Buckle V, Holt S, Munro E, Hunter D, Craig I. Muscular dystrophy in girls with X;autosome translocations. J Med Genet 1986; 23: 484–490
5 Francke U, Ochs HD, de Martinville B et al. Minor Xp21 chromosome deletion in a male associated with expression of Duchenne muscular dystrophy, chronic granulomattous disease, retinitis pigmentosa, and McLeod syndrome. Am J Hum Genet 1985; 37: 250–267
6 Kunkel LM, Monaco AP, Middlesworth W, Ochs HD, Latt SA. Specific cloning of DNA fragments absent from the DNA of a male patient with an X chromosome deletion. Proc Natl Acad Sci USA 1985; 82: 4778–4782

7 Monaco AP, Bertelson CJ, Middlesworth W et al. Detection of deletions spanning the Duchenne muscular dystrophy locus using a tightly linked DNA segment. Nature 1985; 316: 842–845

8 Monaco AP, Bertelson CJ, Colletti-Feener C, Kunkel LM. Localization and cloning of Xp21 deletion breakpoints involved in muscular dystrophy. Hum Genet 1987; 75: 221–227

9 Monaco AP, Neve RL, Colletti-Feener C, Bertelson CJ, Kurnit DM, Kunkel LM. Isolation of candidate cDNAs for portions of the Duchenne muscular dystrophy gene. Nature 1986; 323: 646–650

10 Koenig M, Hoffman EP, Bertelson CJ, Monaco AP, Feener C, Kunkel LM. Complete cloning of the Duchenne muscular dystrophy (DMD) cDNA and preliminary genomic organization of the DMD gene in normal and affected individuals. Cell 1987; 50: 509–517

11 Burmeister M, Monaco AP, Gillard EF, van Ommen GJB, Affara NA, Ferguson-Smith MA, Kunkel LM, Lehrach H. A 10-megabase physical map of human Xp21, including the Duchenne muscular dystrophy gene. Genomics 1988; 2: 189–202

12 Hoffman EP, Monaco AP, Feener CC, Kunkel LM. Conservation of the Duchenne muscular dystrophy gene in mice and humans. Science 1987; 238: 347–350

13 McAfee MB, Kunkel LM. Conservation of the dystrophin gene through evolution. In preparation

14 Lemaire C, Heilig R, Mandel JL. The chicken dystrophin cDNA: Striking conservation of the C-terminal coding and 3' untranslated regions between man and chicken. EMBO J 1988; 7: 4157–4162

15 Moser H. Duchenne muscular dystrophy: pathogenic aspects and genetic prevention. Hun Genet 1984; 66: 17–40

16 Chapman VM, Murawski M, Miller D, Swiatek D. Mouse News Letter 1985; 72: 120

17 Lev AA, Feener CC, Kunkel LM, Brown RHJr. Expression of the Duchenne's muscular dystrophy gene in cultured muscle cells. J Biol Chem 1988; 262: 15817–15820

18 Chamberlain JS, Pearlman JA, Muzny DM, Gibbs RA, Ranier JE, Reeves AA, Caskey CT. Expression of the Duchenne muscular dystrophy gene in muscle and brain. Science 1988; 239: 1416–1418

19 Nudel U, Robzyk K, Yaffe D. Expression of the putative Duchenne muscular dystrophy gene in differentiated myogenic cell cultures and in the brain. Nature 1988; 331: 635–638

20 Chelly J, Kaplan JC, Maire P, Gautron S, Kahn A. Transcription of the dystrophin gene in human muscle and non-muscle tissues. Nature 1988; 333: 858–860

21 Lindolf M, Kaariainen H, van Ommen GJB, de la Chapelle A. Microdeletions in patients with X-linked muscular dystrophy: molecular-clinical correlations. Clin Genet 1988; 33: 131–139

22 Hu X, Burghes AHM, Ray PN, Thompson MW, Murphy EG, Worton RG. Partial gene duplication in Duchenne and Becker muscular dystrophy. J Med Genet 1988; 25: 369–376

23 Monaco AP, Bertelson CJ, Liechti-Gallati S, Moser H, Kunkel LM. An explanation for the phenotypic differences between patients bearing partial deletions of the DMD locus. Genomics 1988; 2: 90–95

24 Koenig M, Beggs AH, Moyer M et al. The molecular basis for Duchenne versus Becker muscular dystrophy: correlation of severity with type of deletion. Submitted

25 Hoffman EP, Fischbeck KH, Brown RH et al. Dystrophin characterization in muscle biopsies from Duchenne and Becker muscular dystrophy patients. N Engl J Med 1988; 318: 1363–1368

26 Malhotra SB, Hart KA, Klamut HJ et al. Frame-shift deletions in patients with Duchenne and Becker muscular dystrophy. Science 1988; 242: 755–759

27 Hoffman EP, Brown RH, Kunkel LM. Dystrophin: the protein product of the Duchenne muscular dystrophy locus. Cell 1987; 51: 919–928

28 Hoffman EP, Hudecki M, Rosenberg P, Pollina C, Kunkel LM. Cell and fiber-type distribution of dystrophin. Neuron 1988; 1: 411–420

29 Hoffman EP, Knudson CM, Campbell KP, Kunkel LM. Subcellular fractionation of dystrophin to the triads of skeletal muscle. Nature 1987; 330: 754–758

30 Knudson CM, Hoffman EP, Kahl SD, Kunkel LM, Campbell KP. Evidence for the association of dystrophin with the transverse tubular system in skeletal muscle. J Bio Chem 1988; 263: 8480–8484

31 Sugita H, Arahata K, Ishiguro T et al. Negative immunostaining of Duchenne muscular dystrophy and mdx muscle surface membrane with antibody against synthetic peptide fragment predicted from DMD cDNA. Proc Jpn Acad 1988; 64: 37–39

32 Zubrzycka-Gaarn EE, Bulman DE, Karpati G et al. The Duchenne muscular dystrophy gene product is localized in the sarcolemma of human skeletal muscle fibres. Nature 1988; 333: 466–469

33 Bonilla E, Samitt CE, Miranda AF et al. Duchenne muscular dystrophy: deficiency of dystrophin at the muscle cell surface. Cell 1988; 54: 447–452

34 Watkins SC, Hoffman EP, Slayter HS, Kunkel LM. Immunoelectron microscopic localization of dystrophin in myofibres. Nature 1988; 333: 863–866

35 Feener CA, Koenig M, Kunkel LM. Alternative splicing of dystrophin mRNA generates isoforms at the carboxy terminus. Nature 1989; 338: 509–511

36 Koenig M, Monaco AP, Kunkel LM. The complete sequence of dystrophin predicts a rod-shaped cytoskeletal protein. Cell 1988; 53: 219–228

37 Hammond RG Jr. Protein sequence of DMD gene is related to actin-binding domain of alpha-actinin. Cell 1987; 51: 1

38 Mimura N, Asano A. Isolation and characterization of a conserved actin-binding domain from rat hepatic actinogelin, rat skeletal muscle, and chicken gizzard alpha-actinins. J Biol Chem 1986; 261: 10680–10687

39 Hoffman EP, Watkins SC, Slayter HS, Kunkel LM. Detection of a specific isoform of alpha-actinin with antisera directed against dystrophin. J Cell Biol 1989; 108: 503–510

40 Lazarides E. From genes to structural morphogenesis: The genesis and epigenesis of a red blood cell. Cell 1987; 51: 345–356

41 Rowland L. Biochemistry of muscle membranes in Duchenne muscular dystrophy. Muscle Nerve 1980; 3: 3–20

42 Schotland DL, Bonialla E, Wakayama Y. Freeze-fracture studies of muscle plasma membrane in human muscular dystrophy. Acta Neuropathol 1981; 54: 189–197

43 Shibuya S, Wakayama Y. Freeze-fracture studies of myofiber plasma membrane in X chromosome-linked muscular dystrophy (mdx) mice. Acta Neuropathol 1988; 76: 179–184

44 Patel K, Voit T, Dunn MJ, Strong PN, Dubowitz V. Dystrophin and nebulin in the muscular dystrophies. J Neurol Sci 1988; 87: 315–326

45 Cooper BJ, Winand NJ, Stedman H et al. The homologue of the Duchenne locus is defective in X-linked muscular dystrophy of dogs. Nature 1988; 334: 154–156

46 Carpenter JL, Hoffman EP, et al. Feline muscular dystrophy with dytrophin. Submitted

47 Ryder-Cook AS, Sicinski P, Thomas K et al. Localization of the mdx mutation within the mouse dystrophin gene. EMBO J 1988; 7: 3017–3021

48 Coulton GR, Morgan JE, Partridge TA, Sloper JC. The mdx mouse skeletal muscle myopathy: I. A histological, morphometric and biochemical investigation. Neuropathol Appl Neurobiol 1988; 14: 53–70

49 Valentine BA, Cooper BJ, deLahunta A, O'Quinn R, Blue JT. Canine X-linked

muscular dystrophy; An animal model of Duchenne muscular dystrophy: clinical studies. J Neurol Sci 1989 (in press)

50 Bell CD, Conen PE. Histopathologic changes in Duchenne muscular dystrophy. J Neurol Sci 1964; 7: 529–544

51 Partridge TA, Morgan JE, Coulton GR, Hoffman EP, Kunkel LM. Conversion of mdx myofibres from dystrophin-negative to 1-1 positive by injection of normal myoblasts. Nature 1989; 337: 176–179

52 Darras BT, Francke U. Normal human genomic restriction-fragment patterns and polymorphisms revealed by hybridization with the entire dystrophin cDNA. Am J Hum Genet 1988; 43: 612–619

53 Darras BT, Blattner P, Harper JF, Spiro AJ, Alter S, Francke U. Intragenic deletions in 21 Duchenne muscular dystrophy (DMD)/Becker muscular dystrophy (BMD) families studied with the dystrophin cDNA: location of breakpoints on HindIII and Bg1II exon-containing fragment maps, meiotic and mitotic origin of the mutations. Am J Hum Genet 1988; 43: 620–629

54 Bonilla E, Schmidt B, Samitt CE et al. Normal and dystrophin-deficient muscle fibers in carriers of the gene for Duchenne muscular dystrophy. Am J Path 1988; 133: 440–445

55 Arahata K, Ishiura S, Ishiguro T et al. Immunostaining of skeletal and cardiac muscle surface membrane with antibody against Duchenne muscular dystrophy peptide. Nature 1988; 333: 861–863

56 Zatz M, Passos-Bueno MR, Rapaport D. Estimate of the proportion of Duchenne muscular dystrophy with autosomal recessive inheritance. Am J Med Genet 1989 (in press)

57 Hoffman EP, Kunkel LM, Angelini C, Clarke A, Johnson M, Harris JB. Improved diagnosis of Becker muscular dystrophy via dystrophin testing. Submitted

58 Bieber FR, Hoffman EP, Amos J. Dystrophin analysis in Duchenne muscular dystrophy: use in fetal diagnosis and in genetic counseling. Am J Hum Genet 1989 (in press)

59 Watkins SC, Hoffman EP, Slayter H, Kunkel LM. Distribution of dystrophin in heterozygote mdx mice. Muscle Nerve 1989 (in press)

60 Brooke MH, Fenichel GM, Griggs RC et al. Clinical investigations in Duchenne muscular dystrophy: 2. determination of the 'power' of therapeutic trials based on the natural history. Muscle Nerve 1983; 6: 91–103

British Medical Bulletin (1989) Vol. 45, No. 3, pp. 644–658
© The British Council 1989

The *DMD* gene analysed by field inversion gel electrophoresis

J T den Dunnen
E Bakker
G-J B van Ommen
P L Pearson
Department of Human Genetics, Sylvius Laboratories, State University Leiden, Leiden, The Netherlands

Genomic and cDNA probes were used to construct a physical map of the *DMD* region including the 2.3 Mb *DMD* gene. FIGE-analysis allows rapid screening of the complete region using only a few probes, detecting deletions or duplications in over 60% of the patients. The technique is especially powerful in the analysis of carrier females. We have found two mutational hotspots; a minor hotspot located proximally and a major hotspot within a large, centrally located intron. Deletions involving this latter intron were studied using 100 kb of cloned DNA sequences. Although breakpoints are spread over the entire region, 5 are clustered within 3 kb.

Analysis of over 250 BMB/DMD families has underscored the importance of germinal mosaicism as a major diagnostic pitfall. At least 14% of new mutation cases involve germinal mosaicism and this still is a lower estimate, due to small family sizes. Hence, relatives of apparent new mutation patients should be considered to have high risk, and require appropriate counselling.

Duchenne muscular dystrophy (DMD) is a lethal degenerative muscle wasting disorder with a frequency of 1/3500 live male births. Its milder equivalent, Becker muscular dystrophy (BMD),

0007–1420/89/0045–0644/$10.00

is less frequent but is due to defects of the same gene. The use of linked RFLP markers and cDNA probes in Southern and FIGE-analysis now yields > 99% accurate diagnosis in more than 95% of BMD/DMD cases.[1] The study of the basic defect of BMD/DMD has revealed a gene of unprecedented size.[2-4] The gene, located in chromosome band Xp21, encodes a 14 kb mRNA which is translated into a 430 kd muscle protein dubbed 'dystrophin'[5,6] which is found in the plasma membrane.[7,8]

Directly from the start of the investigations which led to the isolation of the *DMD* gene, techniques for the separation of large DNA molecules were used to construct physical maps of the region of interest.[4,9,10] As more and more probes were isolated, the map covered ever larger regions.[2,3,11] Finally, the dystrophin cDNA was isolated.[6,12,13] The map, completed with the 3' end of the cDNA, shows that the *DMD* gene spans about 2.3 Mb of DNA.[11] This physical map was the basis for investigating DNA samples of patients and carrier females from DMD families.[14] Due to the fact that many mutations were caused by deletion or duplication of genomic material, FIGE-analysis proved to be very useful.

FIGE analysis of DMD families was started to address several problems at the same time. First, conventional electrophoresis enables detection of fragments only up to 10–20 kb which limits the detection of rearrangements. Using rare cutting enzymes and long-range electrophoresis, the same genomic probes detect fragments of 100–1000 kb.[14] This enlarges the window within which rearrangements can be detected by one to two orders of magnitude.

Secondly, in families where the index patient is not available, the detection of duplications and deletions in genomic DNA of carrier females is difficult. One has to rely on quantitative comparisons of band intensities of Southern blot results. This is difficult, especially where fragments comigrate or hybridize weakly. In these cases, FIGE-data are more convincing, since a dual signal is obtained: one from the normal and one from the mutated X-chromosome. The latter has an altered size and thus migrates differently.

In the third place, more complex rearrangements like translocations, insertions or inversions may not be visible when genomic or cDNA probes are used in a conventional analysis. Fragments with altered sizes are found only when the event occurs close to the probe used. In the case of translocations, the same probes clearly detect rearrangements in FIGE-analysis.[3,4] Probes flanking the translocation point even give two new signals, one from either side of the translocation. Insertions usually result in larger fragments, while inversions will only be visible when they involve sequences straddling a restriction site.

This paper summarizes the data obtained with both the FIGE- and the conventional analysis. Rearrangements detected by FIGE-analysis were checked using conventional DNA analysis with the dystrophin cDNA to prove that they disrupt expressed sequences. We show a long range map of the DMD region, examples of the FIGE-analysis both in patients and carrier females, the distribution of mutations in relation to the dystrophin cDNA and a specific analysis of the deletion hotspot within a large centrally located intron. Furthermore, we discuss the effects of the recent detection of germline mosaicism[15,16] as a major contribution to the appearance of new mutations in our family material.

METHODS AND MATERIALS

All methods used to digest, electrophorese, blot or hybridize DNA samples have been described extensively.[11,14,17,18] Genomic probes used were 754 = DXS84, pERT84 = DXS142, XJ1 = DXS206, pERT87 = DXS164, JBir = DXS270, P20 = DXS269, GMGX11 = DXS239, J66 = DXS268 and L1 = DXS68. cDNA probes used were cDNA(0–2a), cDNA(0–2) (cDNA-XJ10), cDNA(2b–3a), cDNA(3b–5a), cDNA(5b–7), cDNA(8) and subclones of cDNA(9–14).[6,13]

RESULTS

The principle of megabase DNA electrophoresis is rather simple. A separation of large DNA molecules is achieved by constantly reorientating the molecules during their migration towards the positive electrode. The way in which this reorientation is forced upon the DNA is determined by the electrophoretic separation unit. Several techniques have been described. We used mainly the Pulsed Field Gel Electrophoresis (PFGE)[9,19], Field Inversion Gel Electrophoresis (FIGE)[14,20] and Contour-clamped Homogeneous Electric Field (CHEF) electrophoresis.[21] In the FIGE set-up the DNA migrates for a defined period (switch time) in a forward direction and then, for only part of this time (usually 33%), in the backward direction (180° angle). This process repeats itself continuously for a specified time (run time). In the PFGE and CHEF systems the change in field direction is only 120°. In the CHEF-system, nearly linear DNA migration throughout the entire width of the gel is achieved by a complex construction of the electrophoretic unit, keeping the field homogeneous.[21] In all cases the switch time defines the size range of DNA molecules which is accurately separated.

Considerations

The screening of DNA samples after restriction enzyme digestion and megabase separation may show several types of genetic rearrangements. Depending on the type and extent of the rearrangement, numerous possibilities of abnormally migrating fragments are expected (Fig. 1). To assess the type of rearrangement it is essential to detect the adjacent fragments, preferentially with probes near the next restriction site and away from the mutation. When mutations involve sequences beyond the boundaries of the gene, these bordering probes are mostly not available. In such cases it is not possible to prove the type of rearrangement unequivocally. The availability of the complete cDNA of a gene involved is by itself not enough and in the case of the *DMD* gene (Fig. 2), probes are still required near *Sfi*I sites /B and /J. For *Sfi*I/B site the nearest probe distal, pERT84, is an alternative.

For the analysis of DMD mutations on a megabase scale we have chosen to use the restriction endonuclease *Sfi*I, which recognizes an eight basepair sequence without -CG- dinucleotides. This enzyme gives a clearly detectable set of properly sized fragments throughout the *DMD* region (50–1000 kb). The physical map for *Sfi*I is shown in Figure 2. Due to the presence of some well-defined partially digestible *Sfi*I sites (/B, /E, /G and /H), the rearranged patterns are more complex than outlined in Figure 1. On the other hand, the partially digestible sites give valuable information about orientation and extent of the mutation.

FIGE results

We have screened a randomly selected set of 46 patients, resulting in the detection of 28 (61%) FIGE-rearrangements (Table 1). Some examples are shown in Figure 3a. Patient DL48.1 shows a normal pERT87 hybridization (lane D) and abnormal, but partially identical, JBir and J66 signals (lanes H and L). This pattern

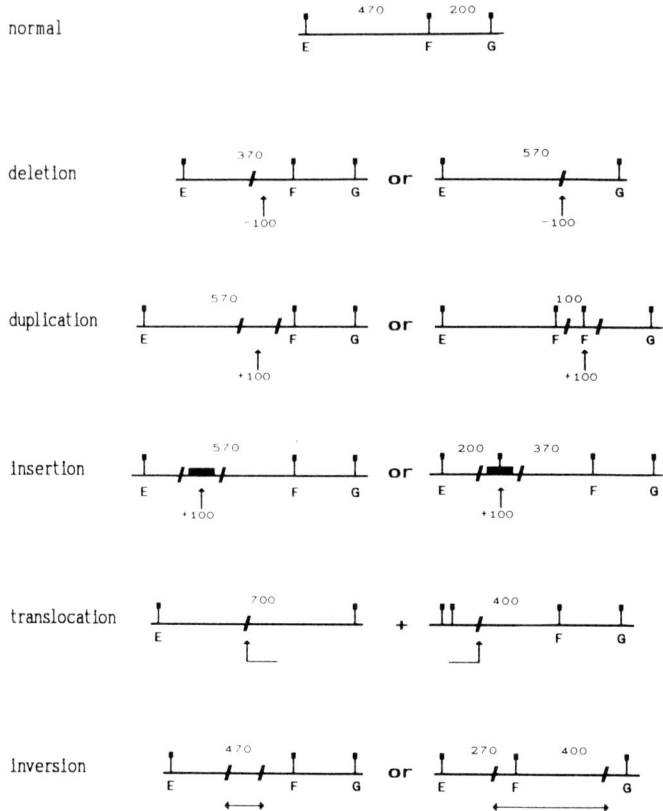

Fig. 1 Schematic representation of possible chromosomal rearrangements and their effect on the physical map. Each mutation, giving a visible effect in a hybridization analysis, is shown individually. Restriction sites are indicated above the map (fragment sizes given in kb) and marked with letters below. Arrows indicate the mutation site while slashes depict their junction sites.

can be explained by a deletion of genomic DNA including the well-digestible *Sfi*I/F site (Fig. 3b). Deletions resulting in altered but simpler hybridization patterns are observed in patients DL23.4 and DL37.5 (Fig. 3). The three deletions mentioned were analysed using the dystrophin cDNA. All three showed deletions of exonic DNA, confirming that the deletions disrupt coding segments of the dystrophin gene.[11]

Sometimes, FIGE-analysis can define an otherwise undetectable anomaly. Figure 3c shows the results of a FIGE-analysis of patient DL185.1. He clearly shows an altered *Sfi*I/B-C fragment, while the flanking 754 and *Sfi*I/C-D hybridization patterns are

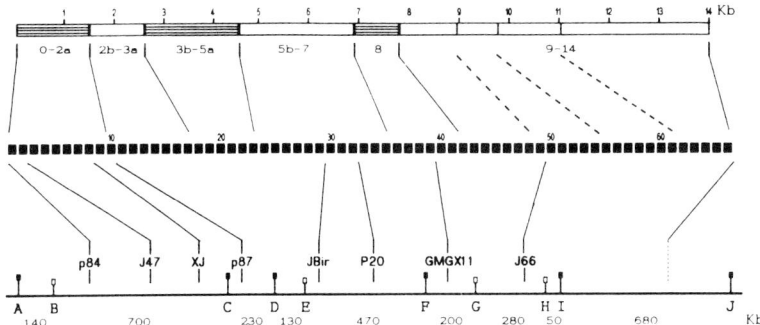

Fig. 2 Schematic view of the dystrophin cDNA in relation to the physical map. Top line; DMD cDNA as it is divided into subprobes for use in hybridization analysis. The numbers indicate size in kb. Central line; the 66 individualy detectable *Hind*III fragments in cDNA hybridizations.[6,11] Bottom line; long-range physical map of the DMD region for the enzymes *Sfi*I.[11] *Sfi*I sites are marked with letters, open boxes show partially digestible *Sfi*I-sites and fragment sizes are given (in kb). The localization of genomic probes is shown. Vertical lines indicate corresponding positions on the three lines.

normal. Analysis with the DMD cDNA, however, revealed no visible alterations.[11] Microscopical analysis of metaphase spreads showed no chromosomal abberations, precluding a visibly detectable translocation. The nature of the mutation in this patient therefore remains to be identified (see DISCUSSION below).

It is evident that FIGE and cDNA data correlate in nearly all cases (Table 1). In 3 cases we were not able to show that the detected rearrangement involved expressed sequences. On the other hand, in four cases where we detected no abnormalities in a FIGE-analysis, cDNA screening did detect alterations. Two patients had a duplication of one exon near the XJ-region, probably involving only a small region of DNA. Two other patients had changes in the normal cDNA hybridization pattern which gave no conclusive answer to the nature of the mutation behind it. Further analysis is required in these cases. Overall, our selected patient set showed at most 32 rearrangements (70%), of which 28 (61%) are detected directly in a FIGE-analysis.

Carrier detection

A powerful area of application of the FIGE technique lies in the analysis of carrier females. Figure 4 shows two examples. DL187.2 carries an alteration in one of her X-chromosomes between the *Sfi*I/D and /E sites. Both mutations are clearly visible and the

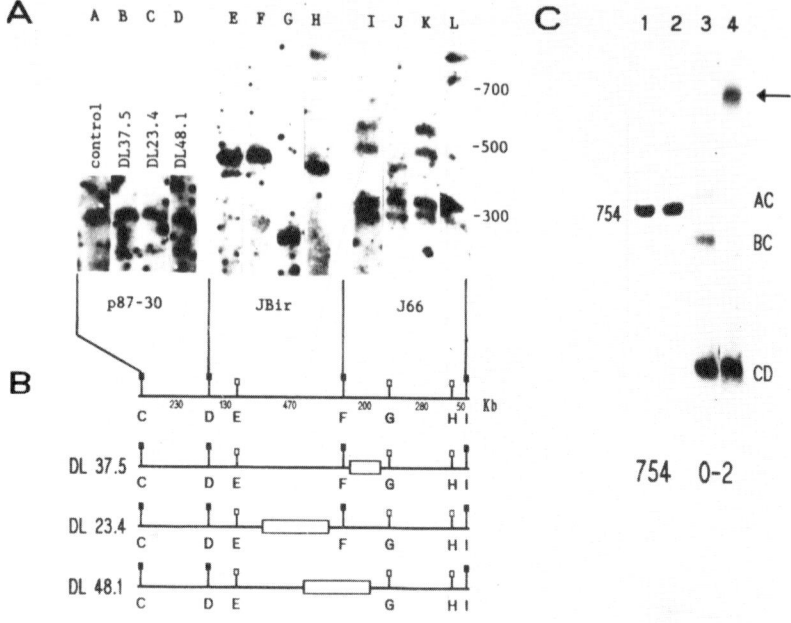

Fig. 3 FIGE-analysis of *Sfi*I digested DNA of several individuals. **A**. Hybridization with DMD probes indicated. Patient DNA's used are given (top of left panel). Fragment sizes are shown on the right. **B**. Location of the deletion detected on the *Sfi*I map. *Sfi*I sites are marked with letters, fragment sizes are given (in kb) and an open box indicates the deletion. **C**. Rearrangement in patient DL185.1 (lanes 2 and 4) next to a control (lanes 1 and 3) as detected after hybridization with the probes indicated (bottom). The fragment lettering is as in Figure 2.

mutated signal is accompanied by the normal signal serving as an internal control, thereby increasing the sensitivity of the analysis.

A comparison of a FIGE- and cDNA analysis, is shown in Figure 4b. Patient DL43.7 has a deletion detectable with cDNA(0–2). This results in cDNA analysis in three missing exonic fragments and in a FIGE-analysis in 60 kb smaller *Sfi*I/B–C and /A–C fragments. The advantage of the FIGE-analysis in carrier detection clearly shows in female DL43.3, who has two easily distinguishable sets of hybridizing FIGE fragments, while in the Southern analysis one has to rely on a quantitative estimate of the hybridizing intensities (Fig. 4b).

Mutation hotspots

When all rearrangements are taken together (Fig. 5) several intriguing features appear.[11] First, the distribution of deletions/

Table 1 FIGE- and cDNA-analysis of DMD patients

*Sfi*I fragment					Fraction		cDNA not confirmed
B–C	C–D	D–F	F–I	I–J	d	%	
+	+	+	+	+	18	39%	−(1)
[]	+	+	+	+	3	7%	2
[	]	+	+	+	3	7%	0
[	−	]	+	+	2	4%	0
+	[	−	]	+	1	2%	0
+	+	[]	+	+	4	9%	0
+	+	[	]	+	7	15%	0
+	+	+	[]	+	6	13%	1
+	+	+	[	−	1	2%	0
+	+	+	+	[]	1	2%	0
Total FIGE detectable mutations					28	61%(2)	3

Overview of the results obtained using FIGE-analysis. Values given (d/N%) show the number of abnormalities detected (d) after screening N patients (N = 46). For FIGE-analysis each type of abnormality detected is shown separately; + for a normal *Sfi*I fragment (Fig. 2), − for a deletion, and square brackets for an altered fragment. The last column shows the number of FIGE rearrangements which could not be confirmed with cDNA hybridizations. (1) Two one exon duplications, not visible in FIGE-analysis, and two unclear cDNA alterations were detected extra. (2) 28/46 or 61%.

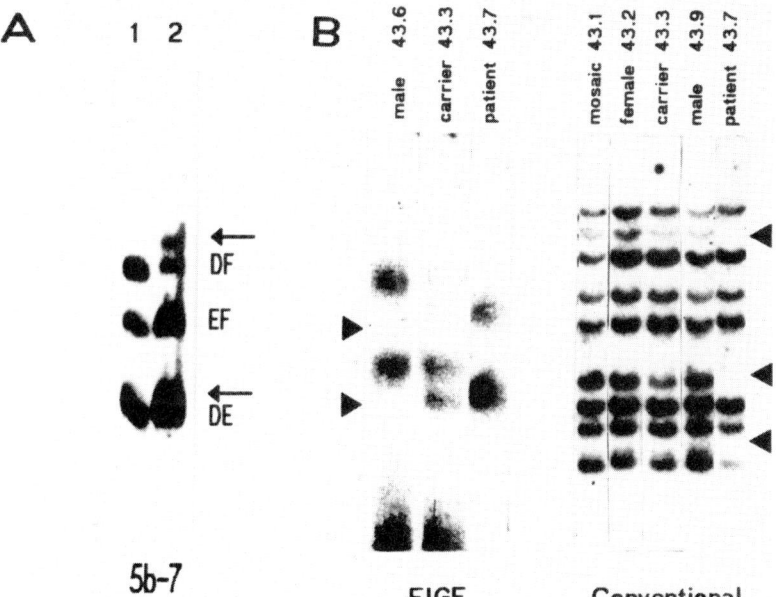

Fig. 4 FIGE-analysis of carrier females. **A.** Carrier female DL187.2 (lane 5) after analysis with the cDNA probe indicated (bottom). Lane 1 shows a control hybridization. **B.** Comparison of FIGE- (left panel) and cDNA-analysis (right panel) in family DL43. The left panel shows a FIGE *Sfi*I-blot and the right panel a conventional *Xmn*I-blot both hybridized with cDNA (0–2). Arrow heads indicate altered (left) or missing (right) fragments. DNA's used are indicated (top).

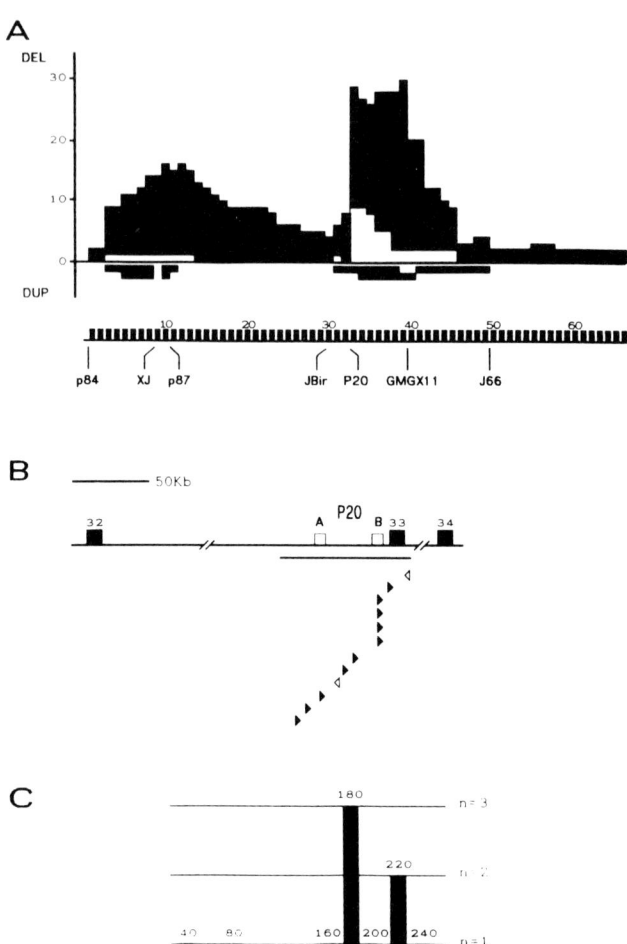

Fig. 5 Mutations in the DMD gene. **A**. Graphical representation of the number of deletions (DEL) and duplications (DUP) detected in the dystrophin gene in relation to exonic HindIII fragments.[6,11] The portion of BMD mutations is shown in white. The localization of genomic probes is indicated below the figure. **B**. Enlargement of the P20 region. Black boxes depict HindIII fragments 32–34 and white boxes the two P20-arms.[18,23] The cloned region is shown as a line below the map. Individually localized deletion breakpoints are marked with arrowheads. **C**. Graphical view of the size of individual deletions in the P20 region as determined in FIGE-analysis. Deletion size is indicated above each bar (grouped in 20 kb size intervals). N determines the number of deletions found per size range.

duplications is not random. A major and a minor mutation 'hotspot' appear, confirming earlier results.[6,14,22] The most pronounced hotspot is located around position 7.0 kb of the cDNA, in a central intron where the sequences defined by probe P20 are located (Fig. 2).[23] The highest proportion of deletions starts in this intron.[11,18] The minor hotspot is located around cDNA position 1.2 kb, surrounding the XJ-intron. Several differences between the two hotspots can be observed:

1. Rearrangements extend into both directions from the minor hotspot, while they have a 3' polarity in the major hotspot.

2. Deletions are very heterogeneous in size in the XJ-hotspot, while a deletion size of around 200 kb is predominant in the major P20-hotspot (Fig. 5c).

3. Two thirds of the duplications are found in the minor hotspot.

4. BMD mutations are markedly clustered in the P20 hotspot.

We analysed mutations in the major mutational hotspot more precisely. Most mutations have breakpoints in the introns between *Hind*III fragments 32 and 34, as detected with the cDNA.[6,11] P20 maps within the intron between Hd32 and Hd33.[23] Through cosmid walking experiments we have cloned nearly 100 kb of overlapping DNA sequences in this region. The cloned region includes both segments of P20 as well as one exon of the *DMD* gene,[18] which is located on the 0.5 kb *Hind*III fragment 33 (Fig. 5b). Twenty-two deletions starting in the P20 intron have been analysed and we have mapped 12 deletion junctions within the cloned region. Although deletion breakpoints are spread over the complete area, 5 lie very close to each other in a 3 kb region (Fig. 5b). Ten deletion breakpoints are located outside the cloned region of intron 32. Collaborative analysis of larger patient sets are in progress, to determine whether distribution of breakpoints elsewhere in this intron is random or confined to several specific sites.

Combining cDNA data, showing which exons are missing, and FIGE data, providing the size of a deletion, results in an overall picture of the exon spacing of the DMD gene over a region of 2.3 Mb of DNA.[11] Deletions around the middle (pERT87–JBir) and in the 3' region (J66) of the gene are rather scarce. The available exon map is thus not complete in those regions.

Germinal mosaicism

Recently, the analysis of DNA families in our and other laboratories revealed another interesting, but at the same time alarming observation.[15,16,28]. We studied 268 DMD families for carrier detection using both flanking and intragenic RFLP probes. Specific analysis of the mutated chromosomes allowed us to deduce the (grand) parental origin of a new mutation in 48 cases. In two apparent 'familial' cases we could show that the deletion mutation was a new mutation as well. In 42 out of 48 cases the mutation was a deletion or duplication, permitting the detection of its emergence and thus giving unequivocal proof of the new mutation. In six pedigrees we observed recurrence of the new mutation, i.e. the presence of germline mosaicism. This suggests that this phenomenon is very frequent. Furthermore, in 14 out of the 42 pedigrees the patient had no other sib carrying the same X-haplotype. Thus, we detected 6 cases of mosaicism in only 28 families. To obtain a first approximation of the frequency of the phenomenon, we have taken as the calculation basis the total number of sibs of the patient carrying the identical X-haplotype. Amongst the 41 at risk X-chromosomes transmitted, 6 were affected. Therefore, the estimated recurrence risk for mothers of a proven new mutation patient becomes 7% for any unspecified male fetus and 14% for the at risk haplotype[28].

DISCUSSION

FIGE-analysis

Chromosomal rearrangements detected by FIGE-analysis were in almost all cases confirmed with conventional Southern blot analyses using the dystrophin cDNA (Table 1).[11] Two mutations were detected which were missed in the FIGE-analysis. This is probably due to the small size of the 'one-exon' duplications involved. In order to detect these small rearrangements the sensitivity of the FIGE-analysis needs to be increased by the use of electrophoretic conditions which optimize separation of DNA molecules in the size range under study. Another means to increase detection is analysis of carrier females since these will result in dual hybridization signals. An advantage of the latter approach is that it compensates for slight variations in migration distances, caused by differences in the amount of DNA loaded per lane, which occasionally hamper lane to lane comparisons, especially in the case of small size differences.

The reproducible, inherited FIGE-rearrangements which could not be confirmed by conventional analysis may involve intronic sequences only, e.g. impeding correct splicing, or regulatory sequences upstream from the gene in the promoter region. Also translocations and inversions may be responsible for alterations detected in a FIGE-analysis without visible changes in cDNA patterns. However, these are relatively seldom and should in most cases result in more complex changes on a megabase scale (Fig. 1). Finally, we can not rule out the alternative possibility that exonic sequences are involved which hybridize poorly or comigrate with other fragments. However, several digestions were done and all fail to highlight anomalies, rendering this possibility less likely. Experiments directed at a more detailed mapping of the chromosomal region involved are needed to clarify these rearrangements. Whatever their nature, they are at least useful markers for the affected X-chromosome in normal haplotype analysis, and most probably highlight the actual mutation itself.

Background of deletion hotspots

The basis for the high mutation frequency of the XJ- and P20-regions remains unclear. In our family material, a 200 kb region surrounding the sequences defined by P20 contains 34 rearrangement sites while the 200 kb region directly upstream, encompassing the sequences defined by JBir, contains only 6 rearrangement sites. This means that either specific sequences or other properties of the region involved, like chromosomal higher order structure, are responsible for this process. When specific recombination sequences exist, many deletions may have breakpoints at or around given specific sites. The involvement of chromosome structure is suggested by the conserved size of deletions around the P20-hotspot. A model accounting for these observations will be presented in a parallel study.[18]

One observation remains puzzling, however. Both mutation hotspots are located in rather large introns, even for the dystrophin gene.[8,11,18] Evolutionary processes are expected to lead to a shortening of these introns, bearing in mind the deletion-prone nature of mutations here. This has not happened, suggesting that essential sequences are located in these introns. A speculative hypothesis is that these essential sequences are other genes, whose expression may not necessarily be linked to *DMD* gene expression. Further studies are required to address the possibility of interdigitation of the *DMD* gene with other genetic units, as was recently reported for the *factor VIII* gene.[24]

Implications of germline mosaicism

Our family screening of over 250 BMD/DMD families has shown the incorrectness of the assumption that detecting the origin of a new mutation in a patient would imply that the mother has no elevated risk in her future pregnancies[15,16,28]. The significant association of *de novo* DMD and BMD mutations with germline mosaicism indicates the need for prenatal diagnosis in subsequent pregnancies.

If the new mutation occurs after meiosis and fertilization, in a mitotic cell division, it will result in a mosaic. The extent of the mosaicism will vary, depending on the embryonic stage at which it has occurred. Early in embryogenesis a mutation will cause a more general somatic mosaicism while later in embryogenesis, in germline proliferation, it might only show up in a fraction of the gametes. These early somatic mutations could account for a number of unexpected side effects. For example, a mild form of muscular dystrophy, which may be misdiagnosed as, e.g. limb-girdle dystrophy, may be the result of somatic mosaicism in a male. Furthermore, in females, an undetected somatic mutation or germline mosaicism introduces a bias in the recombination data for DNA markers relative to the mutation. For example, in family DL43, we have originally scored two recombinants for probe 754 relative to the DMD mutation[17] which contributed significantly to the genetic distance of 3 cM between *DXS84* (754) and *DMD*.[25] This interpretation turned out to be incorrect due to the presence of mosaicism in this family, so that the true genetic distance between *DXS84* and *DMD* has decreased.

Towards a yet better prevention

Recent developments have facilitated carrier detection and prenatal diagnosis in most families, with a continuously increasing reliability. The ultimate goal of detecting the mutation itself, is already possible in about 65% of the cases. Future developments will aim at increasing this number, by increasing the sensitivity of the methods used, by looking at the protein product itself, or by sequencing the mRNA directly to detect point mutations or splicing abnormalities. Further improvements will result from the introduction of the PCR-method,[26] dramatically saving time and labour in both patient and carrier detection. It will aid in direct detection of mutations[18] and normal haplotype analysis. The method avoids the time consuming and laborious techniques of DNA-isolation, digestion, blotting, hybridization, washing and autoradiography. The number of patients will further decrease when new-born population screenings are set up to detect (new) DMD/BMD mutations.[27] The study of the *DMD* gene itself and its protein product will give answers to the question of the underlying structural or metabolic defect. The question whether or not this will ultimately lead to a therapy or even a cure of existing patients can not be answered yet. However, the progress in the study of this enigmatic gene has been enormous. Further steps along the same path are the only rational and thus obligatory course to take.

ACKNOWLEDGEMENTS

We thank Martin Wapenaar, Ieke Ginjaar, Lau Blonden and Henk Veenema for their collaboration and helpful discussions, Erik Bonten, Marleen van Paassen and Petra Grootscholten for expert technical assistance, Ch. van Broeckhoven for providing patient DNAs included in these studies and Drs L Kunkel, R Worton and M Ferguson-Smith for kindly providing the probes used in this analysis. The Dutch Prevention Fund, the Netherlands Organisation for Scientific Research, the Muscular Dystrophy Group of Great Britain and the Muscular Dystrophy Association of the USA are gratefully acknowledged for their generous financial support.

REFERENCES

1 Bakker E, Bonten EJ, Veenema H, et al. Prenatal diagnosis of Duchenne muscular dystrophy: a three year experience in a rapidly evolving field. J Inherit Metab Dis 1989. (in press)
2 Burmeister M, Monaco AP, Gillard EF, et al. A 10 Megabase map of human Xp21, including the Duchenne muscular dystrophy gene. Genomics 1988; 2: 189–202
3 Van Ommen GJB, Bertelson, CE, Ginjaar HB, et al. Long-range genomic map of the Duchenne muscular dystrophy (DMD) gene: Isolation and use of J66 (DXS268), a distal intragenic marker. Genomics 1987; 1: 329–336
4 Kenwrick S, Patterson M, Speer A, Fischbeck K, Davis KE. Molecular analysis of the Duchenne muscular dystrophy region using pulsed field gel electrophoresis. Cell 1987; 48: 351–357
5 Koenig M, Monaco AP, Kunkel LM. The complete sequence of dystrophin predicts a rod-shaped cytoskeletal protein. Cell 1988; 53: 219–228
6 Koenig M, Hoffman EP, Bertelson CJ, Monaco AP, Feener C, Kunkel LM. Complete cloning of the Duchenne muscular dystrophy (DMD) cDNA and preliminary genomic organization of the DMD gene in normal and affected individuals. Cell 1987; 50: 509–517
7 Arahata K, Ishiura S, Ishiguro T, et al. Immunostaining of skeletal and cardiac

muscle surface membrane with antibody against Duchenne muscular dystrophy peptide. Nature 1988; 333: 861–863

8 Zubrzycka-Gaarn EE, Bulman DE, Karpati G, et al. The Duchenne muscular dystrophy gene product is localized in sarcolemma of human skeletal muscle. Nature 1988; 333: 466–469

9 Van Ommen GJB, Verkerk JMH, Hofker, et al. A physical map of 4 million base pairs around the Duchenne muscular dystrophy gene on the human X-chromosome. Cell 1986; 47: 499–504

10 Burmeister M, Lehrach H. Long-range restriction map around the Duchenne muscular dystrophy gene. Nature 1986; 324: 482–485

11 Den Dunnen JT, Grootscholten PM, Bakker E, Van Broeckhoven C, Pearson PL, Van Ommen GJB. Topography of the DMD gene. Manuscript submitted

12 Monaco AP, Neve RL, Colletti-Feener C, Bertelson CJ, Kurnit DM, Kunkel LM. Isolation of candidate cDNAs for portions of the Duchenne muscular dystrophy gene. Nature 1986; 323: 646–650

13 Burghes AHM, Logan C, Hu X, Belfall B, Worton RG, Ray PN. A cDNA clone from the Duchenne/Becker muscular dystrophy gene. Nature 1987; 328: 434–437

14 Den Dunnen JT, Bakker E, Klein-Breteler EG, Pearson PL, Van Ommen GJB. Direct detection of more than 50% Duchenne muscular dystrophy mutations by field inversion gels. Nature 1987; 329: 640–642

15 Darras BT, Francke U. A partial deletion of the muscular dystrophy gene transmitted twice by an unaffected male. Nature 1987; 329: 556–558

16 Bakker E, Van Broeckhoven CH, Bonten EJ, et al. Germline Mosaicism and Duchenne Muscular Dystrophy mutations. Nature 1987; 329: 554–556

17 Bakker E, Bonten EJ, de Lange LF, et al. DNA probe analysis for carrier detection and prenatal diagnosis of Duchenne muscular dystrophy: A standard diagnostic procedure. J Med Genet 1986; 23: 573–580

18 Blonden LAJ, Den Dunnen JT, Van Paassen HMB. High resolution deletion breakpoint mapping in the DMD gene by whole cosmid hybridization. Manuscript submitted

19 Schwartz DC, Cantor CR. Separation of yeast chromosome-size DNAs by pulsed field gradient electrophoresis. Cell 1984; 37: 67–75

20 Carle GR, Frank M, Olson MV. Electrophoretic separations of large DNA molecules by periodic inversion of the electric field. Science 1986; 232: 65–68

21 Chu G, Vollrath D, Davis RW. Separation of large DNA molecules by contour-clamped homogeneous electric fields. Science 1986; 234: 1582–1585

22 Forrest SM, Cross GS, Speer A, Gardner-Medwin D, Burn J, Davies K. Preferential deletion of exons in Duchenne and Becker Muscular Dystrophies. Nature 1987; 329: 638–640

23 Wapenaar MC, Kievits T, Hart KA, et al. A deletion hotspot in the Duchenne muscular dystrophy gene. Genomics 1987; 2: 101–108

24 Levinson B, Lakich D, Silvera P, Kenwrick S, Gitschier J. A gene contained within a factor VIII intron is identified by a CpG island (0765). Am J Hum Genet 1988; 43: A192

25 Hofker MH, Wapenaar MC, Goor N, Bakker E, Van Ommen GJB, Pearson PL. Isolation of probes detecting restriction fragment length polymorphisms from X-chromosome specific libraries: potential use for diagnosis of Duchenne muscular dystrophy. Hum Genet 1985; 70: 148–156

26 Chamberlain JS, Gibbs RA, Ranier JE, Nguyen PN, Caskey CT. Deletion screening of the Duchenne muscular dystrophy locus via multiplex DNA amplification. Nucl Acids Res 1988; 16: 11141–11156

27 Greenberg CR, Rohringer M, Jacobs HK, Averill N, Nylen E, Van Ommen GJB, Wrogemann K. Gene studies in new born males with Duchenne muscular dystrophy detected by Neonatal screening. Lancet 1988; ii: 425–427

28 Bakker E, Veenema H, den Dunnen JT, et al. Germinal mosaicism increases the recurrence risk for 'new' Duchenne muscular dystrophy mutations. J Med Genet (in press)

British Medical Bulletin (1989) Vol. 45, No. 3, pp. 659–680
© The British Council 1989

Molecular analysis of Duchenne and Becker muscular dystrophies

D R Love, S M Forrest, T J Smith, S England, T Flint, K E Davies
Nuffield Department of Clinical Medicine, John Radcliffe Hospital, Oxford, UK

A Speer
Akademie der Wissenschaften der DDR, Zentralinstitut für Molekularbiologie, Abt. Molekular Humangenetik, Berlin-Buch, DDR

The analysis of DNA from patients suffering from Duchenne (DMD) and Becker (BMD) muscular distrophies has resulted in the identification of a single gene locus for these diseases. The locus is deleted to varying extents in affected patients. The translation product of this locus has been implicated as the site of the primary biochemical defect responsible for these muscle disorders.

There is no simple correlation between the severity of the clinical phenotype and the location and extent of genomic deletions in the *DMD* locus. This lack of correlation may be due, in part, to the difficulties inherent in examining a gene of complex arrangement, with exons distributed over a large genomic distance. This paper examines the location of deletion breakpoints in DMD and BMD patients. The molecular analysis of these deletions are presented in the context of transcriptional and translational studies of *DMD* gene expression and the manifestation of the clinical phenotype.

CLINICAL PHENOTYPE

Duchenne muscular distrophy (DMD, McKusick No. 31020) was first reported by Duchenne as a severe debilitating muscle wasting disease affecting boys in early childhood.[1] Clinical signs are generally noticed before the age of 5 years when affected individuals show difficulty in climbing stairs.[2] The proximal muscles

0007–1420/89/0045–0659/$10.00

are affected first with muscle fibres undergoing degeneration and regeneration with extensive connective and adipose tissue proliferation. Approximately 97% of DMD boys are unable to walk and require a wheelchair by the age of 12 years. Most of the affected boys die in their late teens from either pneumonia or heart failure, due to the progressive weakness of respiratory and cardiac muscle, respectively.

Approximately 30% of all cases of DMD exhibit varying degrees of mental retardation.[3,4] This retardation is not progressive and does not correlate with the severity of the disease.

DMD is inherited as an X-linked recessive trait which is transmitted by female carriers and affects males.[2] The incidence of DMD is approximately 1 in 3000 live male births[2] and the disease shows a very high new mutation rate.[5] Thus many of the cases that present in the clinic are new mutations with no previous family history. In view of this, an effective treatment of DMD is an important goal to be reached before any affected families can be helped. The current breakthroughs in the molecular analysis of the *DMD* gene should assist in this aim.

Becker muscular dystrophy (BMD, McKusick no. 31010) is a more benign form of X-linked muscular dystrophy which is characterized by a milder clinical course of proximal muscle weakness.[2,6] Affected boys tend to be confined to wheelchairs much later than DMD patients and in some cases can live a normal life span. An arbitrary cutoff of 12 years of age in becoming wheelchair-bound is often used to distinguish between the clinical presentations of DMD and BMD. However, BMD is much more heterogeneous than DMD. Quite severely affected BMD patients present in their late teens whereas others do not exhibit symptoms until their forties or fifties.

The incidence of BMD at birth is approximately 1 in 30 000 males and patients are rarely mentally retarded.[7] BMD patients have sometimes been diagnosed as suffering from spinal muscular atrophy or limb girdle dystrophy. In some cases, it is only on examination of the gene involved that a precise diagnosis can be made. This is particularly true in isolated cases where there is no informative family history to suggest X chromosome linkage.

There are a few boys whose clinical phenotype is intermediate between DMD and BMD. It has been suggested that these patients represent a third clinically defined group, termed outliers.[8] A typical DMD clinical course may be observed in patients until they are fourteen or fifteen years of age, but this is followed by a slower rate of progression of the disease. The affected individuals may live until their thirties or forties.

Carrier females of DMD and BMD sometimes exhibit clinical manifestation of the disease. These cases are thought to be due to selective X chromosome inactivation in which the active X chromosome in most cells is the one carrying the mutation. Therefore the affected females appear as manifesting heterozygotes. An extreme example of this has been seen in a case of monozygotic twins, one of whom is a gymnast while the other is severely affected by DMD.[9]

There are rare instances in which females are affected by DMD or BMD but have no family history of the disease. Some of these cases have been shown to be due to the presence of a balanced X;autosome translocation. Cytogenetic analysis of the sites of exchange have shown that they lie within the Xp21 region on the short arm of the X chromosome.[10] Generally it is the normal X chromosome which is inactivated and the translocated X chromosome which is active in patients with balanced X;autosome translocations.[11] However the severity of the disease in affected girls has been demonstrated to depend on the proportion of normal X chromosomes that are active.[12]

LOCALIZATION OF THE DMD/BMD GENE

The sites of exchange in Xp21 and the pattern of X chromosome inactivation in translocation females suggested that the translocation disrupted the *DMD* gene.[13] Linkage analysis using DNA markers bridging the Xp21 region confirmed the cytogenetic assignment of the *DMD* gene.[14–16] These markers were used subsequently to show that *BMD* mutations mapped to the same region of Xp21 as those of *DMD* mutations, thus providing strong evidence that these mutations are allelic.[17]

DMD and BMD have been shown to be distinct genetically from Emery-Dreifuss dystrophy (EDMD, McKusick no. 31030) which is an X-linked muscular dystrophy with a benign clinical course.[18] The EDMD locus has been mapped to Xq27-qter in the region of the glucose-6-phosphate dehydrogenase gene on the long arm of the X chromosome.[19–21]

The use of human DMD cDNAs has localized the mouse homologue of *DMD* (*mDMD*) to the human equivalent of Xq26-qter on the mouse X chromosome.[22] An X-linked murine muscular dystrophy mutation, called *mdx*, has also been shown to map to this chromosomal region.[23] However, homozygous *mdx* mice exhibit very mild muscle pathology without the extensive connective tissue proliferation evident in human DMD muscle.[24–26] Recent studies have shown that the *mdx* mutation is located within the *mDMD* gene.[27] Therefore, mutations in the mouse and human *DMD* loci give rise to apparent species-dependent variation in clinical phenotype. A comparison of the transcription and translation products of the *mDMD* and human *DMD* loci will be presented later.

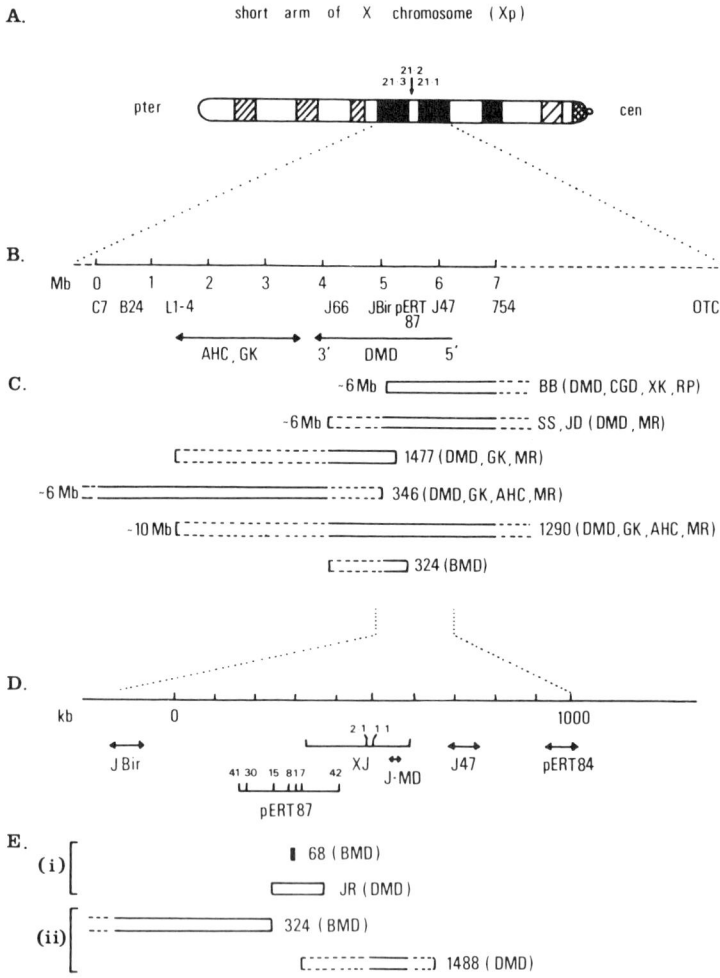

ANALYSIS OF MUTATIONS WITH GENOMIC AND cDNA PROBES

Cloned genomic probes from within the Xp21 region have been used in pulsed field gel studies to construct a long-range restriction enzyme map of Xp21 (Fig. 1). These studies, together with the analysis of affected patients, have shown that the *DMD* locus is approximately 2.3Mb in length.[42,44–47] Gene deletion is a common mechanism of mutation in DMD and BMD. The deletions are highly heterogeneous occurring in different regions across the locus and varying in size. The deletions range from 6000 kb to as little as 6 kb (Fig. 1, patients SS and 68) and there seems to be no correlation between the location and extent of the deletion and the severity of the disease. The manifestation of mental retardation also appears to be totally unrelated to the size or position of the deleted sequences within the *DMD* locus. For example, the related DMD patients SS and JD (see Fig. 1) exhibit a cytogenetically detectable deletion of approximately 6000 kb, but they are discordant for mental retardation; SS is more affected than JD.[30] In

---

Fig. 1 Composite diagram of the extent of DMD and BMD patient deletions relative to a physical map of Xp21.

A. Ideogram of high-resolution CTG bands on Xp (modified from Francke et al[28]). The values above the ideogram designate the location of sub-bands of Xp21.

B. Physical map of human Xp21 (modified from Burmeister et al[29]) indicating the approximate positions of cloned genomic probes. The distance between the loci identified by J66 and L1-4 might be larger than shown; the scale is in megabase pairs (Mb). The dashed line indicates uncertainty regarding the distance between probe 754 and the location of the ornithine transcarbamylase (*OTC*) gene. The extent and direction of transcription of the *DMD* gene are shown as well as the location of the glycerol kinase (*GK*) and congenital adrenal hypoplasia (*AHC*) genes.

C. Open bars indicate the extent of patient deletions relative to the map presented in panel B. The solid lines indicate the known extent of the deletions whereas the dashed lines indicate the limits of uncertainty of the deletion endpoints. The approximate sizes of some of the deletions were determined by the use of flow cytometry.[30] The patients have been reported elsewhere: BB;[28,31] SS;[30] JD;[30] 1477;[32] 346;[32,33] 1290;[32,33] 324.[34,35] The clinical phenotypes of the patients are indicated: DMD, BMD, GK, AHC, MR (mental retardation), CGD (chronic granulomatous disease), XK (McLeod phenotype) and RP (retinitis pigmentosa).

D. Expanded physical map encompassing the region of Xp21 bounded by the genomic probes JBir and pERT84. The arrows indicate the limits of the positions of the probes. The loci identified by pERT87[36,37] and XJ1[38–41] overlap by approximately 70 kilobases (kb). The numerical designations of subclones within these loci are indicated above the horizontal lines denoting the extent of the loci.

E. Open bars indicate the extent of deletions in a selection of DMD and BMD patients. These are drawn as described for panel B. The patients are designated by their original code number or initials: (i)[42] and (ii).[34,35,43]

contrast, patient 1477 (Fig. 1) is deleted for a smaller more distal region of the *DMD* gene compared with that seen in SS and JD, but he is also mentally retarded.[32] The extent of the large genomic deletions found in the DNA of some DMD patients who do not exhibit a complex phenotype (patient JD, Fig. 1) raises the possibility of the existence of other genes flanking the *DMD* locus. The expression of these genes, which may be muscle-specific, or indeed the possibility of alternative transcripts within the *DMD* locus itself, may play some role in the variation of phenotype observed in BMD patients.

Deletions in the proximal region of the DMD gene

The pattern of deletions determined using genomic probes suggests an apparent clustering of 5′ endpoints of deletions in the proximal region of the *DMD* gene. We have studied 15 patients (13 DMD, 2 BMD) who are deleted for the genomic probe HIP25, and all but three have their proximal breakpoints between the loci HH1 and J47.[43] The proximal breakpoints of these 12 deletions have been dissected further by using the 5′ DMD cDNA Cf27 as probe and determining the pattern of exon deletions (Fig. 2).[43] The results indicate that the endpoints do not occur at exactly the same position in the same intron but can be separated by exons. The proximal deletion breakpoints of the 12 patients were divided equally between the introns flanking the 5 kb and 2.5 kb exon-containing *Pst* I fragments (Fig. 2).

Heilig et al.[41] have reported that of 5 independent DMD patients (from 100 screened) who were deleted for a region 30 kb 3′ of XJ1.1, 3 had their 5′ endpoints within a region of 20 kb to the proximal side of JMD (Fig. 2). The significance, if any, of the apparent clustering of deletion breakpoints at the 5′ end of the DMD gene reported by the two groups described above[41,43] (Fig. 2) must await the sequencing of the region and the exact mapping of the various deletions.

Figure 2 illustrates the extent of the deletions observed in two DMD (1489 and 1517) and two BMD (1487 and 1516) patients who are deleted to the proximal side of the genomic probe HIP25.[43] The deletions observed in the two BMD patients remove two extra exons compared with the DMD patients. The reason for the milder clinical phenotype when more exons are deleted is unclear although exon-phasing may account for this apparent paradox; this theory will be discussed later. The two

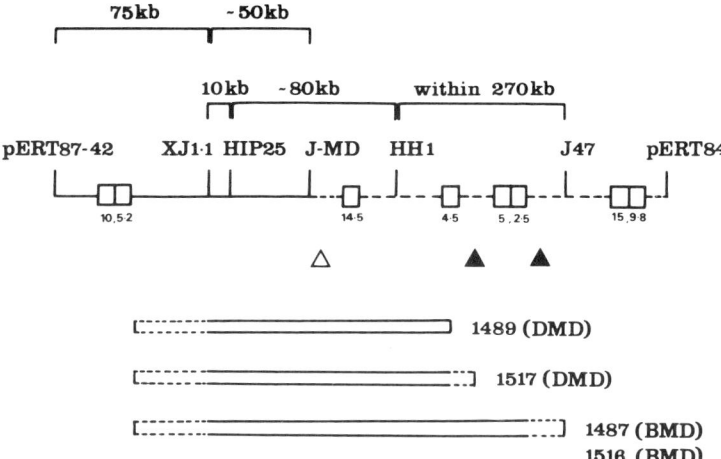

Fig. 2 Localization of deletion endpoints in affected patients in relation to cDNA and partial physical maps of the 5′ end of the DMD gene.

The solid and dashed lines represent distances estimated from restriction enzyme mapping[41] and pulsed field gel studies,[43] respectively.

The open blocks denote exon-containing *Pst* I fragments identified by the 5′ DMD cDNA Cf27.[43,48] The sizes of the *Pst* I fragments are indicated below the open blocks. The exon-containing fragments have been mapped relative to the genomic markers only[43,48] and should not be considered as localizations relative to a physical map.

The open arrow indicates an apparent cluster of deletion breakpoints in 3 DMD patients.[41] This cluster is localized with respect to the physical map only. The dark arrows indicate the apparent clustering of breakpoints in 12 affected patients.[43] These two apparent clusters are localized in relation to genomic and cDNA markers only.

The horizontal open bars denote the location of deletions in 2 DMD (1489 and 1517) and 2 BMD (1487 and 1516) patients.[43] The solid lines indicate the known extent of the deletions whereas the dashed lines indicate the limits of uncertainty of the deletion endpoints.

unrelated BMD patients also illustrate that apparently identical deletions may be manifest as varying phenotypes. Patient 1516 has classical BMD whereas patient 1487 was wheelchair-bound at the age of 11 years.[35] Thus patient 1487 appeared to be following a DMD disease progression. However, further muscle weakness did not develop and he is now running his own business at the age of 35 years. Sequencing the deletion breakpoints of patients 1487 and 1516 may provide some insight into the variation of clinical phenotype.

Deletions in the central region of the DMD gene

The DMD cDNAs have provided a means of examining in more detail the location and extent of deletions compared with genomic probes. Taken together, the cDNAs identify a minimum of 60 exons with a deduced mean size of 200bp for exons and 35 kb for introns.[49] Fetal and adult muscle cDNAs have been sequenced and no differential splicing has been detected over the 5' region of the gene complex.[34,50] The cDNAs can detect deletions in approximately 70% of DMD and BMD patients.[51]

The contiguous cDNAs Cf23a and Cf56a encompass the central 2.5 kb of the DMD cDNA sequence (see later). Together, these cDNAs have been shown to identify deletions in approximately 50% of BMD and DMD patients.[51] Figures 3 and 4 show the hybridization analysis of BMD and DMD patients with Cf23a and Cf56a. The order of the exon-containing fragments identified by the cDNAs has been deduced from the pattern of missing bands in patients with deletions.[52] This analysis assumes that deletions remove contiguous exons and that rearrangements and double deletions are rare.

The characterization of the deletions in 107 DMD and 36 BMD patients who had no previously detectable deletions with 5' DMD probes is shown schematically in Figure 5. All of the BMD patients who exhibited deletions with Cf23a were found to have breakpoints located in the intron separated by exons E and F. Approximately 30% (19/59) of the DMD patients whose deletions were detected by Cf23a and Cf56a had deletion breakpoints which originated in the same intron as that found in BMD patients. This high frequency of deletion breakpoints in one intron has been confirmed by other workers. Koenig et al.[49] reported that 36% of 104 DMD patients had deletion breakpoints which originated within a single intron defined by their probe 7; this probe is the equivalent of Cf23a. The reason why this intron contains the majority of deletion breakpoints is unclear. The intron may span a large genomic distance or contain a sequence-specific hot spot for deletions. The latter possibility would require a detailed analysis of the breakpoints to determine whether they cluster within a small region of the intron.

Most of the BMD patient deletions which were detected with Cf23a were also detected with Cf56a. The majority of the deletions removed exon I only (Fig. 3; common end-point, Fig. 5) with some encompassing exon J as well (patient 1064, Fig. 3). The majority of the DMD patient deletions were detected with Cf56a

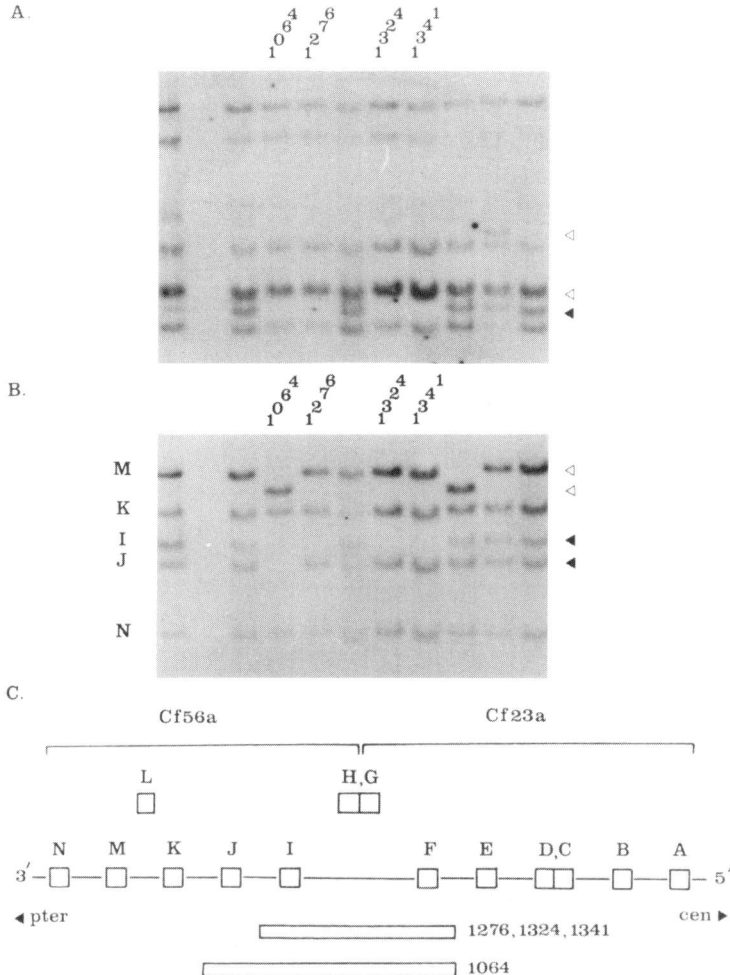

Fig. 3 Hybridization analysis of DNA from BMD and DMD males, digested with *Taq* I.

BMD patients are indicated by their code numbers—track 1: male control; other tracks: DMD patients. Dark arrows indicate the exon-containing fragments which are deleted in the BMD patients. Open arrows indicate *Taq* I restriction fragment length polymorphic (RFLP) bands.

A. Hybridization with Cf23a. The lower of the two open arrows denotes the common RFLP allele which appears as a double dose rather than a single dose.

B. Hybridization with Cf56a. The letter designations for the exon-containing fragments are indicated to the left hand side of the autoradiograph.

C. Schematic diagram of the patient deletion patterns. The three small exons G, H, and L are not considered in this analysis because they are sometimes difficult to detect. The open boxes indicate the location and extent of the deletions relative to the exon map.

rather than Cf23a. These deletions varied both in their size and the region in which they originated. No specific deletion pattern was detected preferentially, in contrast with the apparent homogeneity of deletions observed in BMD patients. None of the deletion breakpoints in DMD patients were found to occur within coding sequence detected by Cf56a.

A complementary study of deletion breakpoints within the central region of the *DMD* gene in affected individuals has been reported by Read et al.[52] (Fig. 6). This study confirmed the clustering of deletion breakpoints of BMD patients between exons E and F, and I and J. The deletions in DMD patients were found to be more heterogeneous by comparison, but with an apparent clustering of breakpoints between exons K and M. In some cases, BMD and DMD patients showed the same apparent deletion pattern. This paradox may be due to the failure to detect very small (< 500bp) exon-containing fragments or exons which do not hybridize well with genomic DNA.

In some families there is variation in the clinical presentation and yet at the DNA level the deletions look similar, if not identical. The pedigree of one of these families is shown in Figure 7. Individuals V–2 and V–4 have classic BMD. They are in their late teens and have difficulty in walking and climbing stairs. They both possess the common BMD deletion of the central portion of the gene (Fig. 5). Their great-great uncle II–2 (patient 1340) also possesses a deletion in the same position of the gene and yet showed no symptoms of the disease until his seventies.[53] He worked as a coalminer and bricklayer at one stage in his life and was still able to manouvre himself into bed until his eighties. This individual provides evidence that it is possible to have a deletion of the coding region of the *DMD* gene and be relatively unaffected by the disease. The sequencing of these patients' breakpoints may not exclude the possibility of point mutation difference. A single base mutation in an intron might introduce an alternative RNA splice site thereby affecting the subsequent translation product. Alternatively, factors other than mutations in the *DMD* locus may serve to effect discordant phenotypes within a pedigree.

TRANSCRIPTION AND TRANSLATION STUDIES OF THE DMD GENE

The DMD cDNAs identify a 14 kb transcript in Northern blots of human fetal and adult skeletal muscle RNA.[49,54,55] A similar sized

A.

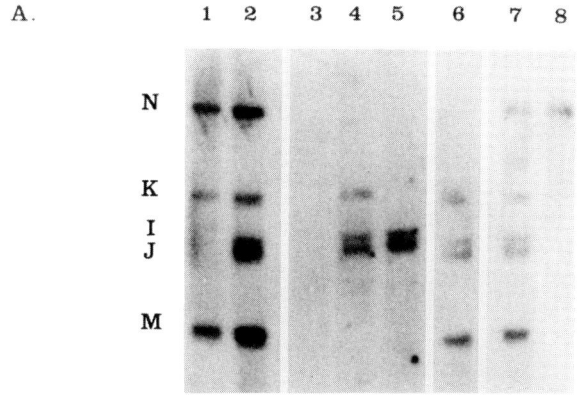

B.

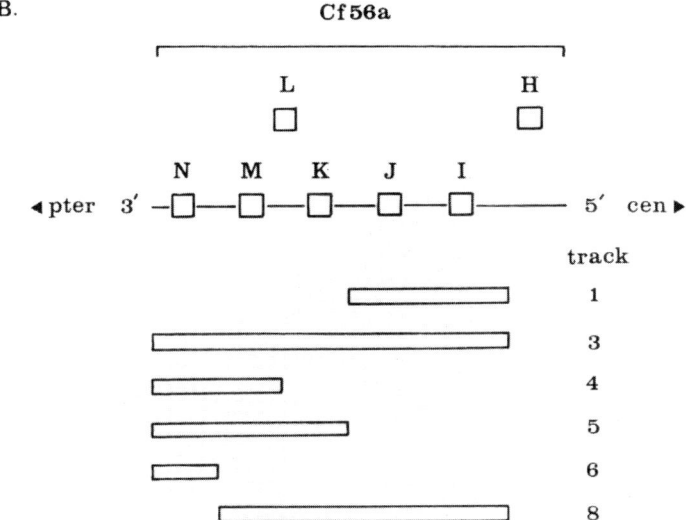

Fig. 4 Hybridization analysis of Pst I-digested DNA from DMD males using Cf56a.
A. The letter designations for the exon-containing fragments are indicated to the left hand side of the autoradiograph.
B. Schematic diagram of the patient deletion patterns deduced from the hybridization results presented in panel A. The diagram is drawn as described in the legend to Figure 3; the small exons H and L are not considered in this analysis.

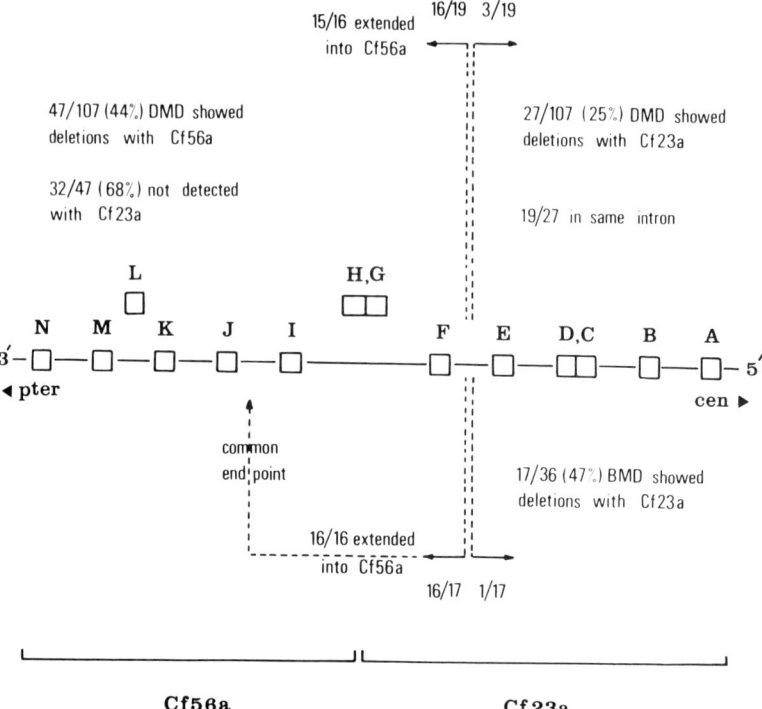

Fig. 5 Summary of the results obtained from screening 107 DMD and 36 BMD patients for deletions using Cf23a and Cf56a.
These results have been reported by Forrest et al.[51] The exon map is described in the legend to Figure 3. The values above the exon map refer to the data obtained from screening the DMD patients while the values below the map refer to the BMD patients' data.

transcript has also been identified in mouse skeletal muscle.[56,57] The human DMD and mDMD cDNAs exhibit approximately 90% sequence similarity over the region for which they have been compared (Fig. 8).[49,56]

Oronzi-Scott et al.[54] used a 1.4 kb cDNA fragment from the 5′ end of the *DMD* gene to examine the expression of the gene in muscle from DMD patients. The *in situ* hybridization analysis of sections of biopsied muscle samples showed a clearly detectable signal in the diseased muscle. However, full-length DMD transcripts were not observed in Northern blots of DMD muscle RNA. These data suggest that the transcription product of the DMD gene in these patients may be truncated or rapidly degraded; no results have been published in a similar analysis of transcripts in

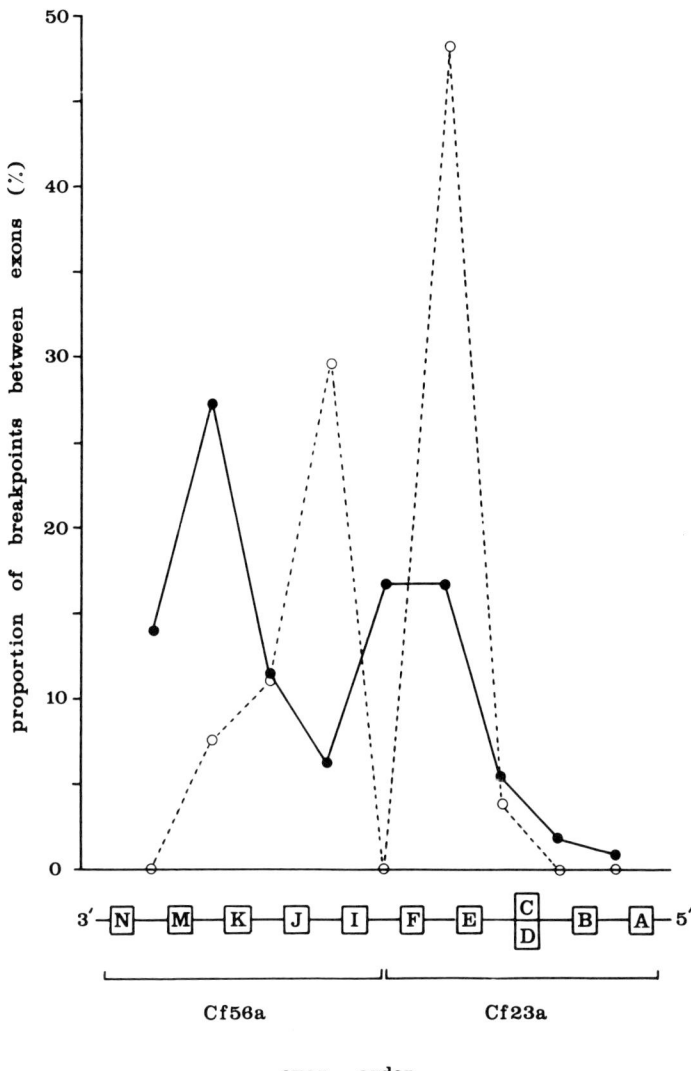

Fig. 6 Analysis of the deletion breakpoints of DMD and BMD patients identified with Cf23a and Cf56a.

A selection of 73 DMD and 14 BMD patients were screened for deletions using Cf23a and Cf56a. Deletion breakpoints were localized with respect to the exon-containing *Pst* I fragments identified by these cDNAs. The filled circles connected by a solid line refer to the DMD patients' results while the open circles connected by a dashed line refer to the results obtained from screening the BMD patients. This analysis is based on the data reported by Read et al.[52]

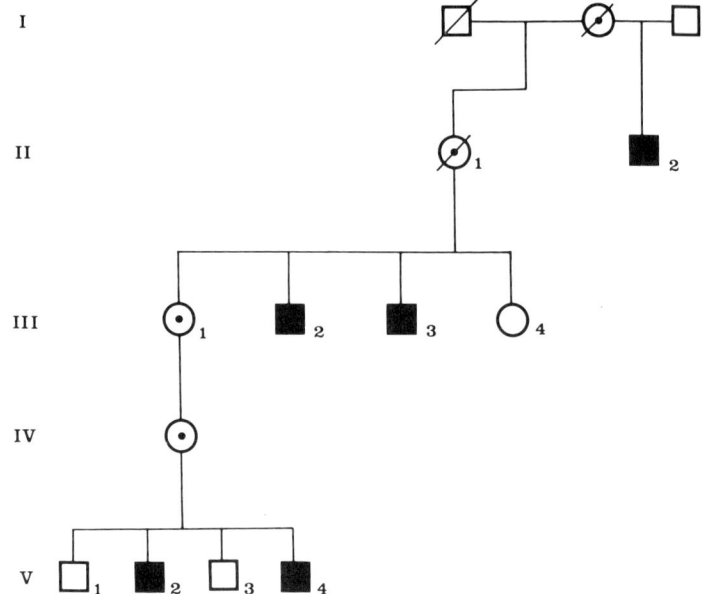

Fig. 7 Pedigree of BMD patient 1340 (II-2).
Open square: male; open circle: female; filled square: affected male; open circle containing an inner filled circle: carrier female; diagonal lines indicate deceased

BMD patients. In the case of *mdx* mice, mRNA of 14 kb has been detected but at a level approximately 16% of that found in normal mice.[57]

The transcription studies have been complemented to a large extent by analysis of the translation product of the *DMD* gene. Hoffman et al.[58] raised antibodies against fusion proteins encoded by two different 5' regions of the *mDMD* gene (Fig. 8) in order to localize the protein product of the human and mouse *DMD* loci. The polyclonal antibodies identified a protein of M_r 400 000 (400 kD) in Western blots of normal human and mouse skeletal and cardiac muscle. This protein has been called dystrophin.[58] Immunolocalization of dystrophin in ultra-thin cryosections of biopsied muscle have shown that the protein is associated with the cytoplasmic face of the sarcolemma.[59] The periodicity of membrane labelling has led to the suggestion of a network of dystrophin molecules which may provide mechanical strength for the plasma membrane.[59-61]

The immunostaining of DMD muscle sections has shown dystrophin to be virtually absent in relation to normal con-

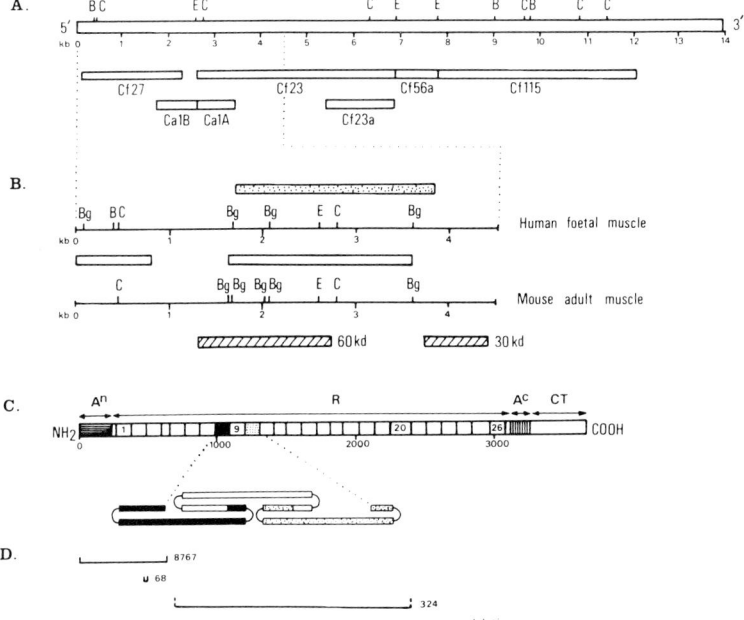

Fig. 8 Schematic illustration of DMD cDNA and deduced domain organization of the translation product of the DMD gene.

A. Partial restriction enzyme map of DMD cDNA taken from Koenig et al.[49] The restriction enzyme sites for *Bam* HI(B), *Eco* RI(E) and *Hin*d II(C) are indicated. The scale of the map is in kb. The open boxes below the map indicate the location and extent of cDNAs isolated from human fetal (Cf) and adult (Ca) muscle cDNA libraries.[34,51]

B. Expanded restriction enzyme maps of the 5' 4.5 kb of human fetal muscle and adult mouse muscle DMD cDNAs.[49,56] The restriction enzyme sites for *Bam* HI(B), *Bgl* II(Bg), *Eco* RI(E) and *Hin*d II(C) are indicated. The stippled box indicates the region over which the human fetal muscle DMD cDNA sequence has been compared with that of the human adult muscle DMD cDNA.[34,50] The open boxes between the restriction enzyme maps indicate the regions over which the human fetal and mouse adult muscle cDNA sequences have been compared.[49,56] The shaded boxes indicate the regions of the mouse adult muscle DMD cDNA which were used to raise antibodies.[58]

C. Organization of protein domains deduced from the DNA sequence of the DMD cDNA.[64] The scale of the domain map is in amino acids, which correlates with the cDNA map presented in panel A. A^n and A^c refer to domains with homology to the amino-terminal and carboxyterminal domains of *Dictyostelium discoideum* α-actinin, respectively.[64] A^n has been proposed to function as an actin-binding domain and A^c is a cysteine-rich domain. R refers to the domain of 26 tandem repeat sequences; the repeats 1,9,20 and 26 are indicated. The proposed tertiary structure of repeats 8 (dark box), 9 (open box) and 10 (stippled box) is shown below the map of domain organization. The boxes represent alpha helices which have been proposed to adopt a tertiary configuration of triple helical structures.[64] CT indicates the carboxyterminal domain.

D. Location and extent of BMD patient deletions relative to the proposed domains of the translation product of the DMD gene (panel C) and the DMD cDNA map (panel A). The patients are designated by their original code numbers: 8767,[64] 68[42,63,64] and 324.[34,35] The common BMD deletion has been presented in Figure 5.

trols.[59-61] This result was confirmed in a recent study by Hoffman et al[62] in which the abundance and size of dystrophin in muscle biopsy specimens from patients with various neuromuscular disorders were determined by Western blotting analysis. The results showed that 35 of 38 DMD patients expressed little or no detectable dystrophin. The milder BMD patients expressed approximately normal levels of abnormal-sized dystrophin. Those patients which were clinically assessed as outliers, referred to earlier, showed variation in both the abundance and size of dystrophin. This study, however, did not examine the characteristics of dystrophin in familial cases with discordant phenotypes. It remains to be seen if these related males have mutations which alter their relative levels of synthesis of dystrophin.

Homozygous *mdx* mouse muscle exhibits no obvious *in situ* immunolabelling using the dystrophin polyclonal antisera.[59,60] This observation contrasts with the expression of the *mDMD* gene at the level of transcription, described above. Therefore, the 14 kb transcript in *mdx* mouse muscle may contain a very small deletion or point mutation such that translation results in a severely truncated or labile protein, or one which is unable to fold into a configuration which expresses the epitopes recognized by the antibodies. The *mdx* mouse presents a paradox when compared with the genetic and phenotypic profiles observed in DMD patients. The consequences of the mutations at the *DMD* loci in both species are apparently the same at the translational level; no detectable dystrophin. However, the clinical phenotypes are quite different as *mdx* mouse muscle does not exhibit the extensive fibrosis found in DMD muscle. Hoffman et al.[58] have suggested that the lack of fibrosis may enable *mdx* muscle to continue regenerating. The different clinical phenotypes may be species-dependent or other factors, as yet unidentified, may play a role in the pathogenesis of this genetic disease.

COMPARISON OF DYSTROPHIN DELETIONS AND CLINICAL PHENOTYPE

Recent studies by Monaco et al.[63] analysed the breakpoints of partial intragenic deletions of the *DMD* locus in a selection of DMD and BMD patients. The deletions observed in the DMD patients were shown to disrupt the open reading frame (ORF) of the mRNA such that translation stop codons were located immediately downstream of the 3' deletion-endpoints. The consequence of these frame-shift mutations would be truncated dystrophin molecules, which may be susceptible to proteolytic degradation. This prediction accords with the Western blot analysis of dystrophin in DMD muscle[62] described above. In the five BMD patients which were examined by Monaco et al.[63], the exons immediately flanking the deletions were shown to maintain an ORF such that a smaller dystrophin molecule would be expected to be produced. The maintenance of an ORF in the context of deletions in the *DMD* locus is termed exon-phasing. Again, the size variation of dystrophin observed in Western blots of BMD muscle[62] is in agreement with the prediction of Monaco et al,[63] based on the small number of patients examined. It does not follow, however, that exon-phasing will be manifest as a BMD phenotype. In view of the apparent functional, albeit truncated, dystrophin produced by BMD patients, these cases offer the opportunity to identify functionally important domains of dystrophin.

The complete nucleotide sequence of the dystrophin transcript has been determined and the deduced amino acid sequence has been used to predict four contiguous domains.[64] The organization of these domains is presented schematically in Figure 8 and are arranged in the following order: an amino-terminal domain of 200 amino acids which has been proposed to function in actin-binding; a middle domain of 26 repeats arranged in tandem with a predicted secondary structure of 25 triple helical segments adopting a rod shape with an approximate length of 125 nm; a cysteine-rich domain which shows homology to the third domain of *Dictyostelium discoideum* α-actinin; and a carboxyterminal domain with no apparent homology to other published sequences.

The mapping of in-frame deletions in several BMD patients, relative to the predicted structure of dystrophin (Fig. 8), indicates that the deletions are located in the two amino-terminal domains.[63,64] The deletion of a significant portion of the proposed actin-binding domain deduced from patient 8767 raises questions about the functional importance of this region of dystrophin. In addition, regions of the tandem repeat sequences appear to be dispensable. The common BMD deletion observed in our patients (Figs 5 and 8) encompasses the 20th repeat sequence and would be predicted to produce a protein approximately 95% of normal size. None of the BMD patient deletions we have analyzed extend close to the 3′ end of the gene. This is also true of DMD deletions, apart from some patients suffering from complex phenotype. Therefore, the deletion patterns which are seen might be due to the mechanism of generation of these deletions rather than one domain being more important than another.

BMD patient 324 is an interesting case because he serves to highlight the fact that large deletions (approximately 5 kb of the dystrophin transcript) can be manifest as a mild phenotype. His deletion removes approximately 60% of the repeat sequence domain (Figs 1 and 8). He showed no clinical symptoms until his late thirties and by his sixties was having difficulty walking.[35] However, his second cousin complained of difficulty in rising from a squatting position and experienced occasional difficulty in walking at the age of 21 years. These symptoms were minor and he only attended a neurologist because of his family history of BMD. Both males have apparently identical deletions. These patients must either produce a deleted dystrophin which is to some extent functional, or have different compensatory mechanisms which reduce the muscle weakness in patient 324 or enhance it in his affected relative.

FUTURE PROSPECTS

The detailed analysis of mildly affected BMD patients relative to severely affected DMD patients should lead to a better understanding of the relative function of the different domains of dystrophin. For example, in vitro-engineered constructs containing selected dystrophin domains could be introduced into muscle cell lines or mdx mice in an attempt to complement the lack of dystrophin. Mouse germ line manipulations would provide an experimental gene therapy in treating muscular dystrophy. However, recent studies in the surgical implantation of autogenic and allogenic myoblasts from primary cultures into neonatal mouse muscle[65-67] raise the possibility of effecting changes in DMD muscle through genetically-engineered somatic cells. Furthermore, if a compensating mechanism exists in some patients exhibiting a mild clinical phenotype then the elucidation of this mechanism might lead to effective strategies for treating the more severe phenotype exhibited by others.

ACKNOWLEDGEMENTS

We would like to thank Helen Blaber for typing this manuscript. We are indebted to the Muscular Dystrophy Group of Great Britain, the Medical Research Council and the Muscular Dystrophy Association of the USA for financial support.

REFERENCES

1 Duchenne GB. Recherches sur la paralysie musculaire pseudohypertrophique ou paralysie myosclérosique. Arch Gen Med (6 ser) 1868; II: 5–25, 179–209, 305–321, 421–443, 552–588

2 Emery AEH, Skinner R. Clinical studies in benign (Becker type) X-linked muscular dystrophy. Clin Genet 1976; 10: 189–201

3 Dubowitz V. Mental retardation in Duchenne muscular dystrophy. In: Rowland LP, ed. Pathogenesis of human muscular dystrophies. Amsterdam: Excerpta Medica, 1977; 688–698

4 Karagan N. Intellectual functioning in Duchenne muscular dystrophy: a review. Psychol Bull 1979; 86: 250–259

5 Moser H. Review of studies on the proportion and origin of new mutants in Duchenne muscular dystrophy. In: ten Kate LP, Pearson PL, Stadhouders AM, eds. Current clinical practice series 20. Amsterdam: Excerpta Medica, 1984; 41–52

6 Becker PE. Eine neue X-chromosomale Muskeldystrophie. Arch Psychiat Neurol Scand 1955; 193: 427

7 Gardner-Medwin D. Clinical features and classification of muscular dystrophies. Br Med Bull 1980; 36: 109–115

8 Brooke MH, Fenichel GM, Griggs RC, et al. Clinical investigations in Duchenne dystrophy: 2 Determination of the 'power' of therapeutic trials based on the natural history. Muscle Nerve 1983; 6: 91–103

9 Burn J, Povey S, Boyd Y, et al. Duchenne muscular dystrophy in one of monozygotic twin girls. J Med Genet 1986; 23: 494–500

10 Boyd Y, Buckle VJ. Cytogenetic heterogeneity of translocations associated with Duchenne muscular dystrophy. Clin Genet 1986; 29: 108–115

11 Leisti JT, Kaback MM, Rimoin DL. Human X-autosome translocations: Differential inactivation of the X chromosome in a kindred with an X-9 translocation. Am J Hum Genet 1975; 27: 441–453

12 Verellen-Dumoulin Ch, Freund M, DeMeyer R, et al. Expression of an X-linked muscular dystrophy in a female due to translocation involving Xp21 and non-random inactivation of the normal X chromosome. Hum Genet 1984; 67: 115–119

13 Boyd Y, Munro E, Ray P, et al. Molecular heterogeneity of translocations associated with muscular dystrophy. Clin Genet 1987; 31: 265–272

14 Murray JM, Davies KE, Harper PS, Meredith L, Mueller CR, Williamson R. Linkage relationship of a cloned DNA sequence on the short arm of the X chromosome to Duchenne muscular dystrophy. Nature 1982; 300: 69–71

15 Davies KE, Pearson PL, Harper PS, et al. Linkage analysis of two cloned DNA sequences flanking the Duchenne muscular dystrophy locus on the short arm of the human X chromosome. Nucl Acids Res 1983; 11: 2303–2312

16 Hofker MH, Wapenaar MC, Goor N, Bakker E, van Ommen G–JB, Pearson PL. Isolation of probes detecting restriction fragment length polymorphisms from X chromosome-specific libraries: potential use for diagnosis of Duchenne muscular dystrophy. Hum Genet 1985; 70: 148–156

17 Kingston HM, Harper PS, Pearson PL, Davies KE, Williamson R, Page D. Localisation of gene for Becker muscular dystrophy. Lancet 1985; II: 1200

18 Emery AEH, Dreifuss FE. Unusual type of benign X-linked muscular dystrophy. J Neurol Neurosurg Psychiatr 1966; 29: 338–342

19 Thomas NST, Williams H, Elsas LJ, Hopkins LC, Sarfarazi M, Harper PS. Localisation of the gene for Emery-Dreifuss muscular dystrophy to the distal long arm of the X chromosome. J Med Genet 1986; 23: 596–598

20 Hodgson S, Boswinkel E, Cole C, et al. A linkage study of Emery-Dreifuss muscular dystrophy. Hum Genet 1986; 74: 409–416

21 Yates JRW, Affara NA, Jamieson DM, et al. Emery-Dreifuss muscular

dystrophy: localisation to Xq27.3-qter confirmed by linkage to the factor VIII gene. J Med Genet 1986; 23: 587–590

22 Heilig R, Lemaire C, Mandel J-L, Dandolo L, Amar L, Avner P. Localization of the region homologous to the Duchenne muscular dystrophy locus on the mouse X chromosome. Nature 1987; 328: 168–170

23 Avner P, Amar L, Arnaud D, Hanauer A, Cambrou J. Detailed ordering of markers localising to the Xq26-Xqter region of the human X chromosome by the use of an interspecific *Mus spretus* mouse cross. Proc Natl Acad Sci USA. 1987; 84: 1629–1633

24 Bridges LR. The association of cardiac muscle necrosis and inflammation with the degenerative and persistent myopathy of *mdx* mice. J Neurol Sci 1986; 72: 147–157

25 Tanabe Y, Esaki K, Nomura T. Skeletal muscle pathology in X chromosome-linked muscular dystrophy (*mdx*) mouse. Acta Neuropathol 1986; 69: 91–95

26 Torres LF, Duchen LW. The mutant *mdx*: inherited myopathy in the mouse. Brain 1987; 110: 269–299

27 Ryder-Cook AS, Sicinski P, Thomas K, et al. Localisation of the *mdx* mutation within the mouse dystrophin gene. EMBO J 1988; 7: 3017–3021

28 Franke U, Harper JF, Darras BT, et al. Congenital adrenal hypoplasia, myopathy, and glycerol kinase deficiency: molecular genetic evidence for deletions. Am J Hum Genet 1987; 40: 212–227

29 Burmeister M, Monaco AP, Gillard EF, et al. A 10-megabase physical map of human Xp21, including the Duchenne muscular dystrophy gene. Genomics 1988; 2: 189–202

30 Wilcox DE, Cooke A, Colgan J, et al. Duchenne muscular dystrophy due to familial Xp21 deletion detectable by DNA analysis and flow cytometry. Hum Genet 1986; 73: 175–180

31 Franke U, Ochs HD, de Martinville B, et al. Minor Xp21 chromosome deletion in a male associated with expression of Duchenne muscular dystrophy, chronic granulomatous disease, retinitis pigmentosa, and McLeod syndrome. Am J Hum Genet 1985; 37: 250–267

32 Davies KE, Patterson MN, Kenwrick SJ, et al. Fine mapping of glycerol kinase deficiency and congenital adrenal hypoplasia within Xp21 on the short arm of the X chromosome. Am J Hum Genet 1988; 29: 557–564

33 Dunger DB, Davies KE, Pembrey M et al. Deletion on the X chromosome detected by direct DNA analysis in one of two unrelated boys with glycerol kinase deficiency, adrenal hypoplasia, and Duchenne muscular dystrophy. Lancet 1986;I: 585–587

34 Cross GS, Speer A, Rosenthal A, et al. Deletions of fetal and adult muscle cDNA in Duchenne and Becker muscular dystrophy patients. EMBO J 1987; 6: 3277–3283

35 Davies KE, Smith TJ, Bundey S, et al. Mild and severe muscular dystrophy associated with deletions in Xp21 of the human X chromosome. J Med Genet 1988; 25: 9–13

36 Monaco AP, Neve RL, Colletti-Feener C, Bertelson CJ, Kurnit DM, Kunkel LM. Isolation of candidate cDNAs for portions of the Duchenne muscular dystrophy gene. Nature 1986; 323: 646–650

37 Kunkel LM and co-authors. Analysis of deletions in DNA from patients with Becker and Duchenne muscular dystrophy. Nature 1986; 322: 73–77

38 Ray PN, Belfall B, Duff C et al. Cloning of the breakpoint of an X;21 translocation associated with Duchenne muscular dystrophy. Nature 1985; 318: 672–75

39 Thompson MW, Ray PN, Belfall B et al. Linkage analysis of polymorphisms within the DNA fragment XJ cloned from the breakpoint of an X;21 translocation associated with X linked muscular dystrophy. J Med Genet 1986; 23: 548–555

40 Burghes AHM, Logan C, Hu X, Belfall B, Worton RG, Ray PN. A cDNA clone from the Duchenne/Becker muscular dystrophy gene. Nature 1987; 328: 434–437

41 Heilig R, Lemaire C, Mandel J-L. A 230 kb cosmid walk in the Duchenne muscular dystrophy gene: detection of a conserved sequence and of a possible deletion prone region. Nucl Acids Res 1987; 15: 9129–9142

42 Monaco AP, Kunkel LM. A giant locus for the Duchenne and Becker muscular dystrophy gene. Trends Genet 1987; 3: 33–37

43 Kenwrick SJ, Smith TJ, England S, Collins F, Davies KE. Localisation of the endpoints of deletions in the 5′ region of the Duchenne gene using a sequence isolated by chromosome jumping. Nucl Acids Res 1988; 16: 1305–1317

44 Burmeister M, Lehrach H. Long-range restriction map around the Duchenne muscular dystrophy gene. Nature 1986; 324: 582–585

45 van Ommen G-JB, Verkerk JMH, Hofker MH, et al. A physical map of 4 million bp around the Duchenne muscular dystrophy gene on the human X-chromosome. Cell 1986; 47: 499–504

46 van Ommen G-JB, Bertelson C, Ginjaar HB, et al. Long-range genomic map of the Duchenne muscular dystrophy (DMD) gene: isolation and use of J66 (DXS268), a distal intragenic marker. Genomics 1987; 1: 329–336

47 Kenwrick S, Patterson M, Speer A, Fischbeck K, Davies KE. Molecular analysis of the Duchenne muscular dystrophy region using pulsed field gel electrophoresis. Cell 1987; 48: 351–357

48 Smith TJ, Forrest SM, Cross GS, Davies KE. Duchenne and Becker muscular dystrophy mutations: analysis using 2.6 kb of muscle cDNA from the 5′ end of the gene. Nucl Acids Res 1987; 15: 9761–9769

49 Koenig M, Hoffman EP, Bertelson CJ, Monaco AP, Feener C, Kunkel LM. Complete cloning of the Duchenne muscular dystrophy (DMD) cDNA and preliminary genomic organization of the DMD gene in normal and affected individuals. Cell 1987; 50: 509–517

50 Rosenthal A, Speer A, Billwitz H, Cross GS, Forrest SM, Davies KE. Nucleotide and corresponding amino acid sequence of human adult and fetal cDNA coding for portions of the Duchenne muscular dystrophy (DMD) gene. Biomed Biochim Acta 1988; 47:K13–K15

51 Forrest SM, Cross GS, Flint T, Speer A, Robson KJH, Davies KE. Further studies of gene deletions that cause Duchenne and Becker muscular dystrophies. Genomics 1988; 2: 109–114

52 Read AP, Mountford RC, Forrest SM, Kenwrick SJ, Davies KE, Harris R. Patterns of exon deletions in Duchenne and Becker muscular dystrophy. Hum Genet 1988; 80: 152–156

53 Forrest SM, Cross GS, Speer A, Gardner-Medwin D, Burn J, Davies KE. Preferential deletion of exons in Duchenne and Becker muscular dystrophies. Nature 1987; 329: 638–640

54 Oronzi Scott M, Sylvester JE, Heiman-Patterson T, et al. Duchenne muscular dystrophy gene expression in normal and diseased human muscle. Science 1988; 239: 1418–1420

55 Lev AA, Feener CC, Kunkel KM, Brown RH Jr. Expression of the Duchenne's muscular dystrophy gene in cultured muscle cells. J Biol Chem 1987; 262: 15817–15820

56 Hoffman EP, Monaco AP, Feener CC, Kunkel LM. Conservation of the Duchenne muscular dystrophy gene in mice and humans. Science 1987; 238: 347–350

57 Chamberlain JS, Pearlman JA, Muzny DM, et al. Expression of the murine Duchenne muscular dystrophy gene in muscle and brain. Science 1988; 239: 1416–1418

58 Hoffman EP, Brown RH Jr, Kunkel LM. Dystrophin: the protein product of the Duchenne muscular dystrophy locus. Cell 1987; 51: 919–928

59 Watkins SC, Hoffman EP, Slayter HS, Kunkel LM. Immunoelectron micro-scopic localization of dystrophin in myofibres. Nature 1988; 333: 863–866

60 Bonilla E, Samitt CE, Miranda AF et al. Duchenne muscular dystrophy: deficiency of dystrophin at the muscle cell surface. Cell 1988; 54: 447–452

61 Zubrzycka-Gaarn EE, Bulman DE, Karpati G, et al. The Duchenne muscular dystrophy gene product is localized in sarcolemma of human skeletal muscle. Nature 1988; 333: 466–469

62 Hoffman EP, Fischbeck KH, Brown RH, et al. Characterization of dystrophin in muscle-biopsy specimens from patients with Duchenne's or Becker's muscu-lar dystrophy. N Engl J Med 1988; 318: 1363–1368

63 Monaco AP, Bertelson CJ, Liechti-Gallati S, Moser H, Kunkel LM. An explanation for the phenotypic differences between patients bearing partial deletions of the DMD locus. Genomics 1988; 2: 90–95

64 Koenig M, Monaco AP, Kunkel LM. The complete sequence of dystrophin predicts a rod-shaped cytoskeletal protein. Cell 1988; 53: 219–228

65 Law PK. Beneficial effects of transplanting normal limb-bud mesenchyme into dystrophic mouse muscle. Muscle Nerve 1982; 5: 619–627

66 Watt DJ, Morgan JE, Partridge TA. Use of mononuclear precursor cells to insert allogenic genes into growing mouse muscles. Muscle Nerve 1984; 7: 741–750

67 Watt DJ, Morgan JE, Partridge TA. Long term survival of allografted muscle precursor cells following a limited period of treatment with cyclosporin A. Clin Exp Immunol 1984; 55: 419–426.

British Medical Bulletin (1989) Vol. 45, No. 3, pp. 681–702
© The British Council 1989

Myogenic regulation of dystrophin gene expression

H J Klamut, E E Zubrzycka-Gaarn, D E Bulman, S B Malhotra, S E Bodrug, R G Worton, P N Ray

The Genetics Department and Research Institute, The Hospital for Sick Children and the Departments of Medical Genetics and Medical Biophysics, University of Toronto, Toronto, Ontario, Canada

Muscle-specific transcriptional regulation of DMD gene expression has been inferred from both the histopathology of the disease and, more recently, from the use of cDNA sequences to detect DMD gene transcripts by Northern blot, RNase protection, *in situ* hybridization, and polymerase chain reaction (PCR) analyses. Several muscle-specific genes have been shown to be transcriptionally activated early in myogenesis and a number of *cis*-acting promoter elements required for muscle-specific transcriptional induction have been described. In this report we review our recent progress on studies of the mechanisms underlying myogenic regulation of dystrophin gene expression. Indirect immunofluorescence has been used to demonstrate that dystrophin is present at the muscle cell surface very early in the myogenic program. The cloning and sequencing of the dystrophin gene promoter reveals the presence of pre-defined muscle-specific *cis*-acting promoter elements. Functional assays provide evidence that these upstream sequences are capable of regulating DMD gene expression in a cell- and developmental stage-specific manner.

Tissue-specific transcriptional control of gene expression involves a complex interaction between *cis*-acting sequences such as enhancers and regulatory elements located in the upstream promoter and intragenic regions of the gene and tissue-specific *trans*-acting

0007–1420/89/0045–0681/$10.00

factors which interact with these *cis*-acting sequences to induce positive or negative regulation of gene transcription.[1,2] The differentiation of determined mononucleated myoblasts into multinucleated myotubes in the process of myogenesis requires the coordinate induction of a variety of muscle-specific gene products and the concommitant down-regulation of a number of genes associated with the undifferentiated state. As reviewed by Buckingham (this Issue) in vitro skeletal muscle myogenesis represents an ideal system within which to study the mechanisms underlying muscle-specific transcriptional regulation, as evidenced by the recent identification of a variety of upstream and intragenic regulatory elements involved in the regulation of a number of muscle-specific contractile, intermediate filament, and enzymatic proteins.[3-14]

The successful application of 'reverse genetics' to the cloning of the *Duchenne muscular dystrophy (DMD)* gene has recently led to great advances in our understanding of this most severe and most common form of muscular dystrophy. The isolation of genomic sequences lying within the *DMD* gene locus at band Xp21 on the short arm of the X chromosome,[15,16] and the subsequent isolation and sequencing of cDNA clones corresponding to the *DMD* gene transcript[17-19] has provided the means for the experimental determination of the mechanisms controlling the developmental expression of this gene and for the elucidation of the biochemical nature of the gene product and its role in the pathogenesis of the disease.

In this review, we examine the potential of utilizing myogenic cell cultures to study the developmental expression and biochemical nature of the *DMD* gene product in the light of recent advances in our understanding of dystrophin expression. In addition, we present our initial findings in regard to the elucidation of the *cis*-acting sequences and *trans*-acting factors which regulate the myogenic expression of the *DMD* gene. The results of these studies complement each other well, and demonstrate that: (i) the *DMD* gene product is expressed early in the myogenic process and is localized to the muscle cell surface; (ii) the *DMD* gene promoter region contains regulatory elements found in other muscle-specific genes; and (iii) these upstream promoter sequences are capable of regulating *DMD* gene transcription in a developmental- and cell-specific manner.

DYSTROPHIN EXPRESSION AND LOCALIZATION WITH MYOGENIC DIFFERENTIATION

Clinical and pathophysiological considerations

Duchenne and Becker muscular dystrophy belong to a group of heritable diseases whose most prominent clinical and histopathological characteristics are progressive weakening and degeneration of muscle tissues. At least 10 clinical forms of muscular dystrophy have been described and classified according to the mode of inheritance, the age of onset, the muscle groups involved, and the clinical course.[20] These discrete clinical manifestations suggest that each disease is the result of a distinctive biochemical defect which primarily affects muscle tissue, but which may also affect other tissues to varying degrees.[21] In many, the defect also appears to be selectively expressed within muscle tissue, affecting a definable subset of muscle groups while others are spared.

Duchenne muscular dystrophy (DMD) is the most severe and most common of this group of muscle disorders, affecting 1/3300 live male births. Although histopathological skeletal muscle abnormalities have been detected very early in the developing fetus, clinical onset typically occurs at 3–5 years of age. The disease progresses rapidly and shows a predictable pattern of differential skeletal muscle degeneration. Other affected tissues include cardiac muscle and the smooth muscle of the gastrointestinal tract, both showing histopathological evidence of degeneration. The central nervous system also seems to be affected, as evidenced by lower mean IQ. Death usually occurs due to respiratory complications before the age of 24.[22] Patient analyses utilizing cDNA sequences corresponding to the *DMD* gene locus have confirmed earlier reports[23] that the phenotypically milder and less common Becker muscular dystrophy (BMD) is caused by an allelic variant of the *DMD* gene and have indicated that over 50% of DMD and BMD patients carry a deletion for all or part of the gene.[24,25]

DMD Gene expression in fetal and adult tissues and in myogenic cell culture

The recent availability of cDNA clones corresponding to the *DMD* gene has prompted a number of studies aimed at defining the tissue-specificity and development pattern of *DMD* gene expression. Initial observations using Northern blot analysis[17,18] were complicated by the large size (14 kb) and relative low abundance of the dystrophin transcript. The application of more sensitive techniques, such as *in situ* hybridization,[26] RNase protection,[27,28] and more recently the polymerase chain reaction (PCR)[29] have clearly indicated that the *DMD* gene transcript is most abundant in skeletal and cardiac muscle (0.02–0.1% of total RNA), and is also present at significant levels in smooth muscle and brain tissue (5% and 1% of skeletal muscle levels, respectively).[29] This distribution is in good agreement with the clinical spectrum of tissue involvement in this disease.

In terms of developmental expression, dystrophin transcripts have been detected in both adult and fetal skeletal muscle tissues,[26,27] as well as in differentiated myotubes in culture.[30] Nudel et al.[27] utilized RNase protection analysis to demonstrate that *DMD* gene transcription is initiated as myoblasts begin to differentiate into multinucleated myotubes, a pattern similar to that of a number of other muscle-specific genes studied to date. No evidence has been obtained to suggest that the *DMD* gene is transcribed in undifferentiated mononucleated myoblasts. Dystrophin mRNA was also undetectable or extremely rare in fibroblasts, HeLa cells and cultured lymphoblastoid cells.[17,18,29] These results indicate, therefore, that *DMD* gene transcription is regulated in a tissue- and developmental stage-specific manner, and is activated during the process of myoblast differentiation.

At the translational level, fusion protein- and peptide-generated antibodies to the amino terminal region of dystrophin have been independently prepared by a number of laboratories and have been used to examine the tissue specificity and subcellular localization of the *DMD* gene product. Hoffman et al.[31,32] have used western blot analysis to demonstrate that dystrophin is expressed predominantly in skeletal, cardiac and smooth muscle of both humans and mice, and at equivalent levels in both adult and fetal tissues. In contrast to observed levels of dystrophin mRNA, Hoffman et al.[32] have reported skeletal and cardiac muscle-equivalent levels of dystrophin in visceral and vascular smooth muscle. They have also noted a slightly lower apparent molecular weight of smooth muscle-derived dystrophin in SDS polyacrylamide gels, and have suggested that this tissue expresses a different isoform of the protein. Somewhat lower levels of dystrophin were observed in brain cortex and spinal cord extracts, and a comparison of dystrophin levels in cultures containing different proportions of neuronal and glial cells indicated that dystrophin is primarily expressed in neuronal cells. Subcellular fractionation was used in association with western blot analysis to demonstrate that dystrophin co-fractionates with heavy sarcoplasmic reticulum, suggesting an intracellular localization at the triadic junction in skeletal muscle.[33] In an extension of these studies, Knudson et al.[34] demonstrated that dystrophin is specifically associated with the junctional transverse tubular membranes.

Immunohistochemical studies in our laboratory[35] and in others[36-38] have indicated that dystrophin is predominantly localized to the skeletal muscle sarcolemma. No staining of intracellular components was observed in either cross- or longitudinal-sections at the light microscopy level, and no preferential distribution to any particular fiber type could be discerned. Watkins et al.[38] confirmed that dystrophin is primarily localized to the cytoplasmic face of the skeletal muscle sarcolemma using immuno-gold electron microscopy of ultra-thin cryosections. A low level of intracellular staining was also observed in the region of the A/I junction of the sarcomere, suggesting an association of dystrophin with the contiguous t-tubular system as well. In addition to skeletal muscle, immunohistochemical analyses have also detected dystrophin in adult cardiac muscle[37] and myocardiocytes.[36] In contrast to the results of Hoffman et al.[32] no dystrophin-specific staining was observed in skin smooth muscle cells[36] or in the walls of capillaries (vascular smooth muscle) in skeletal muscle sections.[37]

Immunolocalization of dystrophin during myogenic differentiation

Elucidation of the biochemical function of dystrophin and its role in the pathogenesis of the disease will require the use of a model system within which a variety of biochemical and genetic manipulations can be readily made. The demonstration that the *DMD* gene is transcribed in differentiating myogenic cultures provides evidence that muscle cell cultures represent such a potential model system. The application of cell culture techniques to the study of myogenic disorders offers the advantage of direct manipulation of specific cell types and their environment free of the complicating influences which exist in the intact tissue. This is particularly important in studies of the pathogenesis of the muscular dystrophies, in which a myriad of secondary events arising from the initiation of muscle degeneration have effectively masked the nature of the primary causative biochemical defect.

Primary muscle cell cultures are readily prepared by enzymatic dissociation of fresh human muscle biopsies.[39] Contamination of primary muscle cell cultures to varying degrees by non-myogenic cells such as fibroblasts can be circumvented by the use of clonal myogenic cell cultures prepared by dilution plating and harvesting of isolated colonies arising from single cells. This approach would seem ideally suited to the study of myoblasts from DMD patients because of the increased proportion of connective tissue and fat in these muscle biopsies. In practice, however, clonal culturing of DMD patient myoblasts has proven difficult because of the depletion in their numbers and their reduced growth potential,[40] presumably as a consequence of the high degree of regenerative activity which occurs in DMD muscle.

We have studied the cell-specific and developmental regulation of dystrophin expression in normal human clonal myoblasts, primary cultures of DMD myoblasts, and normal human fibroblast cultures. Dystrophin was detected with affinity-purified polyclonal rabbit antibodies prepared against a protein A fusion protein (F927) and a BSA-conjugated peptide (P929) containing sequences corresponding to portions of the amino-terminal end of human dystrophin. The preparation of antibodies and their affinity purification has been described in detail elsewhere.[35] For these experiments, cells were seeded at a density of 50 cells/mm^2 on multi-well tissue culture slides in growth medium (Alpha MEM containing 16 mM glucose, 10% fetal bovine serum and 40 μg/ml gentamycin) and allowed to proliferate for 3–5 days until confluent. Myoblasts were allowed to fuse into multinucleated myotubes in the presence of fusion medium (growth medium containing 1–2% fetal bovine serum). At appropriate stages of growth and differentiation, cells were washed in PBS, fixed and permeabilized with absolute methanol for 1 minute and air dried. Cells were incubated in the presence of the first antibody (1:30 dilution in PBS) for 3 hours at room temperature in a humidified chamber, washed 3 times for 10 minutes each with PBS containing 0.1% Tween-20, and exposed for 1 hour at room temperature to either fluorescein- or rhodamine-conjugated goat anti-rabbit antibody

Fig. 1 Immunofluorescence staining of human cell cultures using peptide- and fusion protein-generated antibodies. Cells were analyzed for affinity purified dystrophin antibody staining by indirect immunofluorescence with rhodamine-conjugated goat anti-rabbit antibodies. Phase contrast (a,c,e,g,i) and corresponding fluorescence micrographs (b,d,f,h,j) of normal human myotubes stained with preimmune serum (a and b), F927 (c and d) and P929 (e and f), DMD myotubes (g and h) and normal human fibroblasts (i and j) stained with P929. Bar = 40 μm

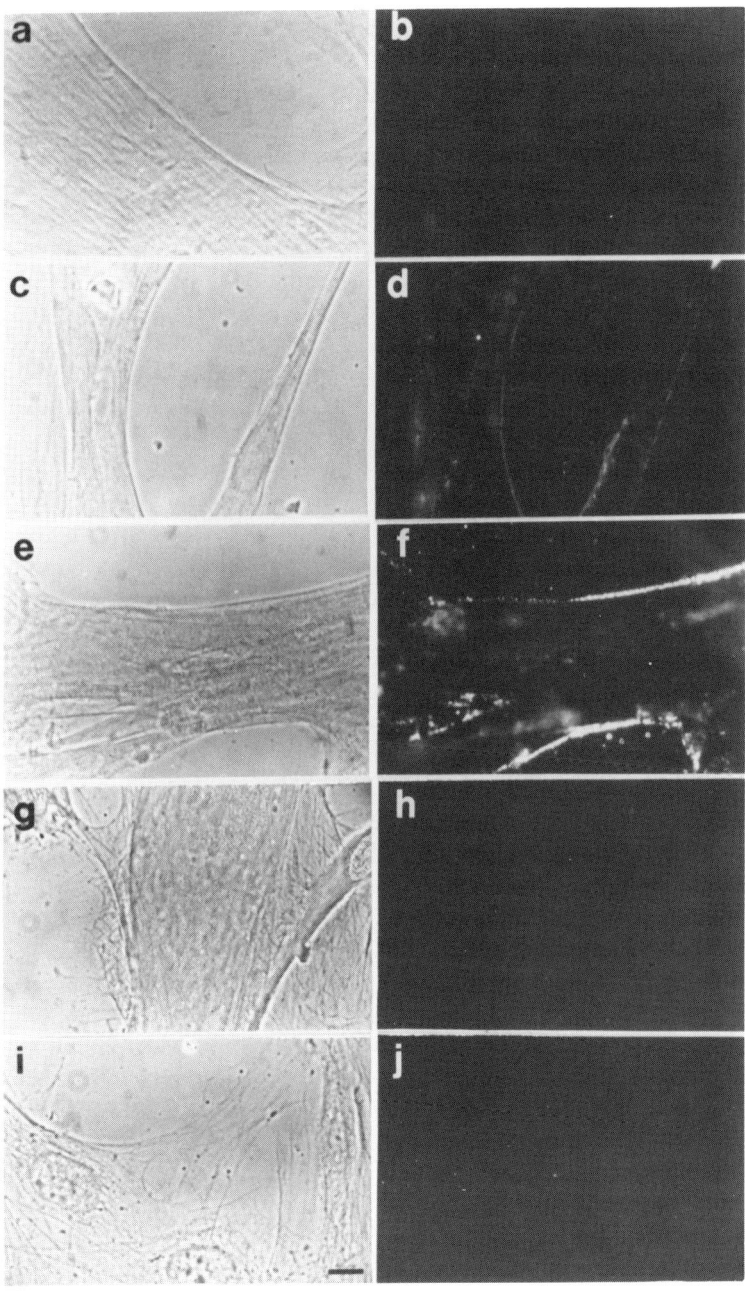

(1:50 dilution). Cells were again washed 3 times for 10 minutes each in PBS containing 0.1% Tween-20 and mounted under 50% glycerol in PBS containing 0.02% sodium azide. Immunolabelled cells were examined and photographed using a Leitz Diaplan microscope equipped with epi-illumination and a 100 × oil objective. The results are illustrated in Figures 1 and 2.

As shown in Figure 1, immunofluorescence staining of dystrophin was consistently observed at the surface of normal clonal differentiated myotubes with either the F927 fusion protein antibodies (Fig. 1d) or the P929 peptide antibodies (Fig. 1f). Staining at the myotube surface occurred in discrete patches and was often seen to follow the longitudinal plane of the myotube. In addition, staining was often found associated with extracellular membrane vesicles (blebs) which are generated as a consequence of the fixation procedure (Fig. 1f), indicating that dystrophin is tightly associated with the surface membrane. No surface staining was observed with preimmune serum (Fig. 1b) or in myotubes from a DMD patient (Fig. 1h) or normal human fibroblasts (Fig. 1j).

A study of dystrophin staining as a function of the fusion process in clonal human myoblasts is presented in Figure 2. Very low levels of dystrophin were evident in predifferentiated mononucleated myoblasts (Fig. 2b), but staining became more prominent in early fusing myoblasts (Fig. 2d) and progressively accumulated at the cell surface as myotubes continued to differentiate and mature (Fig. 2f, 2h). Some nuclear staining was also evident at the early stages of myoblast differentiation (Fig. 2d), but disappeared in the more mature myotube. This series of experiments suggests that dystrophin synthesis is initiated very early in myogenesis and accumulates steadily at the cell surface as myoblasts differentiate and mature into multinucleated myotubes, in agreement with the pattern of *DMD* gene transcription observed by Nudel et al.[27] Perhaps more significant, however, is the demonstration that dystrophin can be detected at the surface of a single myogenic cell grown in vitro, an observation that has broad implications in terms of the applicability of this system to studies of the biochemical role of dystrophin in the normal muscle cell.

REGULATION OF DYSTROPHIN GENE EXPRESSION

Control of myogenic gene expression

The determination and differentiation of tissue-specific cell lineages requires the coordinate induction and repression of a large

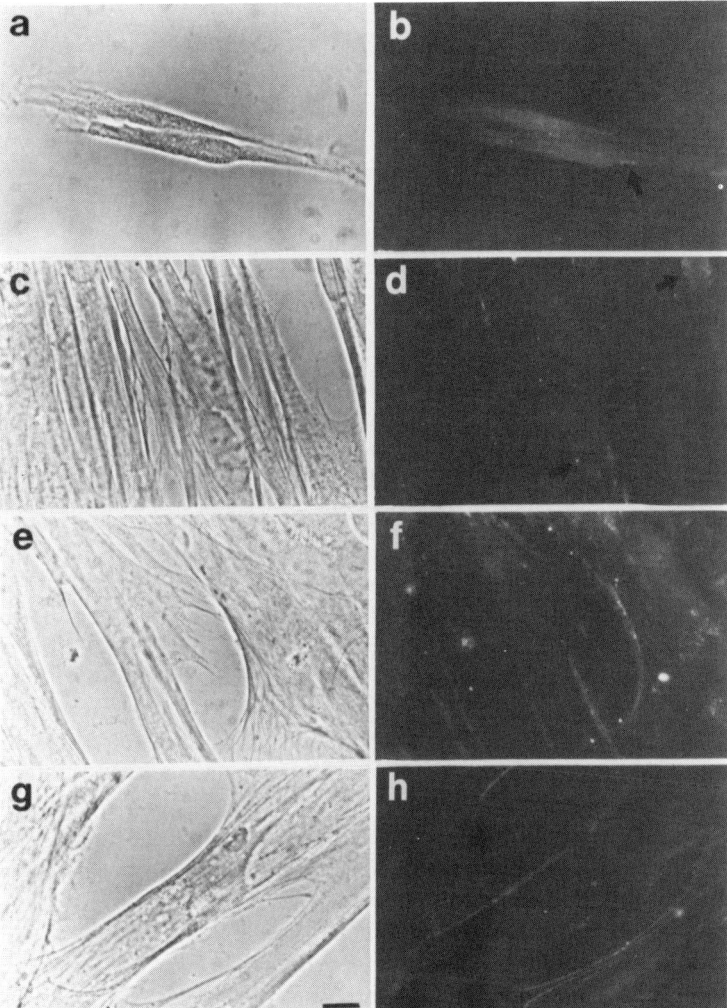

Fig. 2 Expression of dystrophin during myogenesis. Normal human clonal myoblasts were stained with affinity-purified P929 rabbit anti-dystrophin antibodies and fluorescein-conjugated goat anti-rabbit antibodies. Phase and fluorescence micrographs of low density pre-fusion myoblasts (day 1) (a and b); near-confluent myoblasts in the early stages of fusion (day 4) (c and d); confluent myoblasts undergoing extensive fusion (day 6); late fully-differentiated myotubes (day 9). Surface-specific dystrophin staining is evident at very low levels in myoblasts (arrow in panel b) and increased steadily throughout the process of differentiation. Some nuclear staining is evident in myoblasts in the early stages of differentiation (arrows in panel d). Bar = 40 μm

number of functionally and genetically independent genes. Evidence has recently been presented that supports the concept that gene switching in myogenic differentiation is controlled by the activation of a relatively small number of specific genetic elements (*trans*-acting factors),[41] that interact with conserved *cis*-acting sequences located in the promoter regions of muscle-specific genes to coordinately activate their transcription.[1]

Generally, the promoter regions of tissue-specific genes are organized such that a non-tissue-specific basal element containing a TATA box (required for correct initiation at the transcription start or CAP site) and one or more CAAT and/or GC boxes (which play a role in controlling the frequency of initiation) is located within the first 50–80 bp upstream of the transcription initiation site. Tissue-specific transcriptional regulation is acquired through the association of one or more unique regulatory elements or enhancers with the basal element, which can be positioned in the near or far upstream regions or, occasionally, within or at the extreme 3' end of the gene.[2]

In order to begin to identify and understand the mechanism by which these regulatory elements interact to control myogenic gene expression, the upstream regions of a variety of muscle-specific genes have been analyzed and a number of potential muscle-specific regulatory elements have been described. The best characterized of these is the CArG box $(CC(A + T\text{-rich})_6 GG)$ identified by Kedes et al. in their analysis of the cardiac and skeletal alpha-actin gene promoters.[3,4] These sequences, present in multiple copies within approximately 200 bp of the transcription initiation site, were shown by deletion and mutational analysis to be required for the muscle-specific activation of these genes. CArG elements were subsequently found to be present upstream of a number of other muscle-specific genes such as the chicken myosin light chain 2, chicken cardiac troponin T, and the mouse muscle creatine kinase genes.[5-7] The recent demonstration that these sequences interact with nuclear factors (CArG box binding factors or CBF) present in both muscle and non-muscle cell types, and that this interaction can be blocked by the human *c-fos* serum response element (which also contains a CArG-like sequence) suggests that additional *cis*-acting sequences and/or *trans*-acting factors are involved in the muscle-specific regulation of these genes.[8,9] Grichnik et al.[10] have recently identified two CBAR (CAAT box-associated repeat) elements symmetrically oriented in the promoter region of the chicken skeletal alpha actin gene which are capable of directing orientation-independent muscle-specific transcription. Interestingly, the CBAR consensus contains within it the CArG box elements described by Kedes et al. In addition, a CBAR element located in the promoter region of the chicken smooth muscle alpha actin gene has been shown to be important in the repression of transcription of this gene in myoblast cultures.[11]

A second muscle-specific promoter element which has recently gained some attention is the MCAT (muscle-CAT) heptameric consensus. Mar et al.[12] have shown that disruption of one or two of these sequence elements within the chicken cardiac troponin T promoter abolishes its ability to direct muscle-specific transcription. Interestingly, deletion of a CArG box present upstream of these MCAT elements in the cardiac troponin T promoter has only modest effects on transcription in muscle cells. The MCAT consensus has also been identified in the promoter regions of a variety of other muscle contractile protein genes, but has not as yet been characterized in regard to its potential interaction with muscle-specific *trans*-acting factors.

As yet, no clear picture has emerged regarding the mechanisms involved in the transcriptional regulation of muscle-specific gene expression. The identification of a relatively small number of conserved sequences within the promoter regions of the muscle-specific gene studied to date supports the notion that the induction and/or repression of a small number of *trans*-acting factors precipitates a cascade of myogenic gene expression. On the other hand, observations of ubiquitous factor binding and dual functionality suggest that the interplay between these factors and their binding sites is much more complex than originally anticipated.

Isolation and characterization of the DMD gene promoter

The recent availability of cDNA sequences from the *DMD* gene and the demonstration that transcription of this gene is induced upon myogenic differentiation has provided the opportunity to isolate and characterize the regulatory sequences governing the tissue- and developmental-specific transcription of this gene. The identification of the regulatory sequences controlling the myogenic expression of this gene should: (i) provide additional insights into the general mechanism of muscle-specific gene regulation, and perhaps aid in the elucidation of dystrophin function by comparison of its regulatory sequences and patterns of expression to other muscle-specific gene families; (ii) allow for the identification of *trans*-acting factors which control the expression of this gene, and which may potentially be involved in the pathogenesis of other muscle diseases; (iii) aid in the identification of potential promoter mutations in DMD patients; and (iv) provide valuable information regarding the potential for the future application of cell and gene replacement therapy to this disease.

Our approach to the cloning of the 5′ end of the gene has been to utilize a 2 kb cDNA clone containing the first 16 exons of the *DMD* gene[18] to screen a human genomic cosmid library. Five cosmid clones containing the first 7 exons of the gene were isolated, extending the XJ walk region (*DXS206* locus) by over 150 kb in the centromeric direction. The most centromeric of these clones (designated XJC8) was found to contain exon 1 as well as sequences complimentary to the *DXS142* (PERT84) locus (Fig. 3). Koenig et al.[24] have also reported that exon 1 lies within the *DXS142* locus. A restriction map of cosmid XJC8 maps exon 1 to a 3.8 kb *Hind* III fragment, a 5.6 kb *Eco* R1 fragment, and a 16.9 kb *Pst*1 fragment, in agreement with Southern blot analysis of genomic digests probed with 5′ cDNA clones.[18,42] A 1.4 kb *Hind* III-*Pst* 1 fragment (designated HP2; Fig. 3) containing exon 1 was subcloned from XJC8 and used to probe Southern blots of human genomic DNA digested with a variety of restriction enzymes. The correlation between band sizes observed in XJC8 digests probed with cDNA and genomic digests probed with HP2 indicated that the integrity of the cloned region surrounding exon 1 had been maintained.

Sequence analysis of HP2 indicated that this fragment consists of 238 bp corresponding to exon 1268 bp corresponding to a portion of intron 1, and 887 bp of upstream sequence representing the potential *DMD* gene promoter region. The complete sequence is shown in Figure 4. The 5′ and 3′ ends of exon 1 were determined by comparison of cDNA and genomic sequences. Since the transcription initiation site (5′ end of exon 1) has not yet been confirmed, the position of the 5′ end of the transcription unit (+ 1 in the sequence of Fig. 4) must be considered tentative. Examination of sequences upstream of exon 1 reveals the presence of a potential basal promoter element consisting of an AT-rich region (TATA box) and a GC box (Sp1 factor binding site) located 61 and 98 bp upstream of the 5′ end of exon 1, respectively (Fig. 4). Since the TATA box is invariably found approximately 30 bp upstream of the CAP site, the true transcription initiation site for

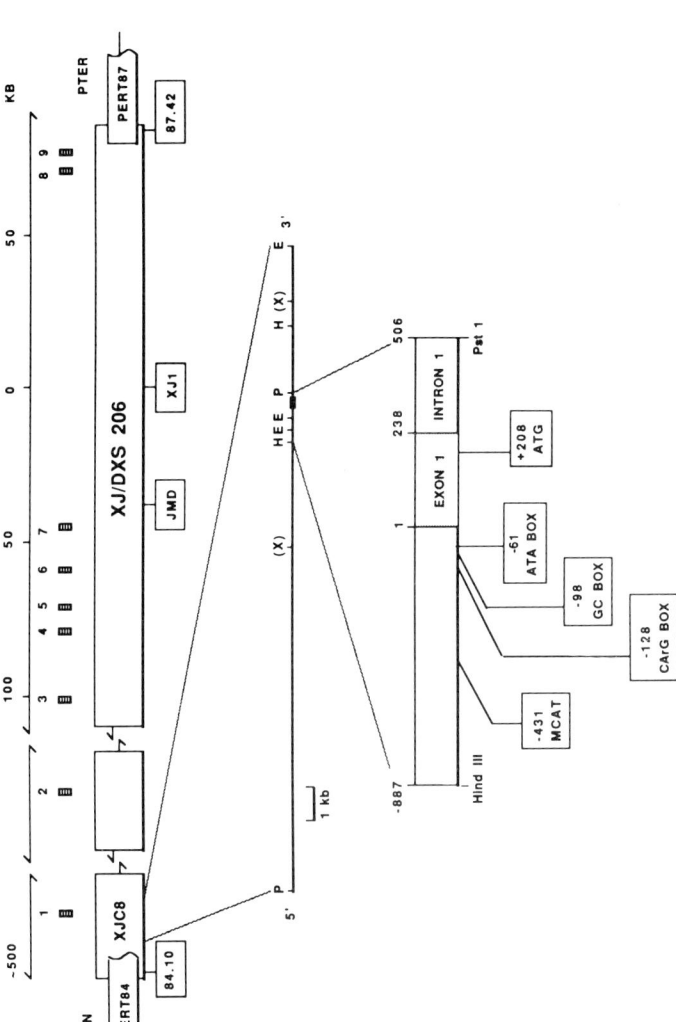

Fig. 3 Isolation and molecular analysis of the *DMD* gene promoter region. The position of cosmid XJC8 containing exon 1 relative to the PERT 84 and XJ/DXS 206 regions of the *DMD* gene. The approximate positions of the first 9 exons within the XJ locus are also shown. An expanded view of the region surrounding exon 1 is shown below, with the position of exon 1 indicated by the solid bar. The 1.4 kb *Hind* III-*Pst* 1 fragment (designated HP2) containing exon 1, a portion of intron 1, and 887 bp of upstream sequence is also shown, along with the location of the predicted ATG translational start site and the positions of potentially relevant promoter elements.

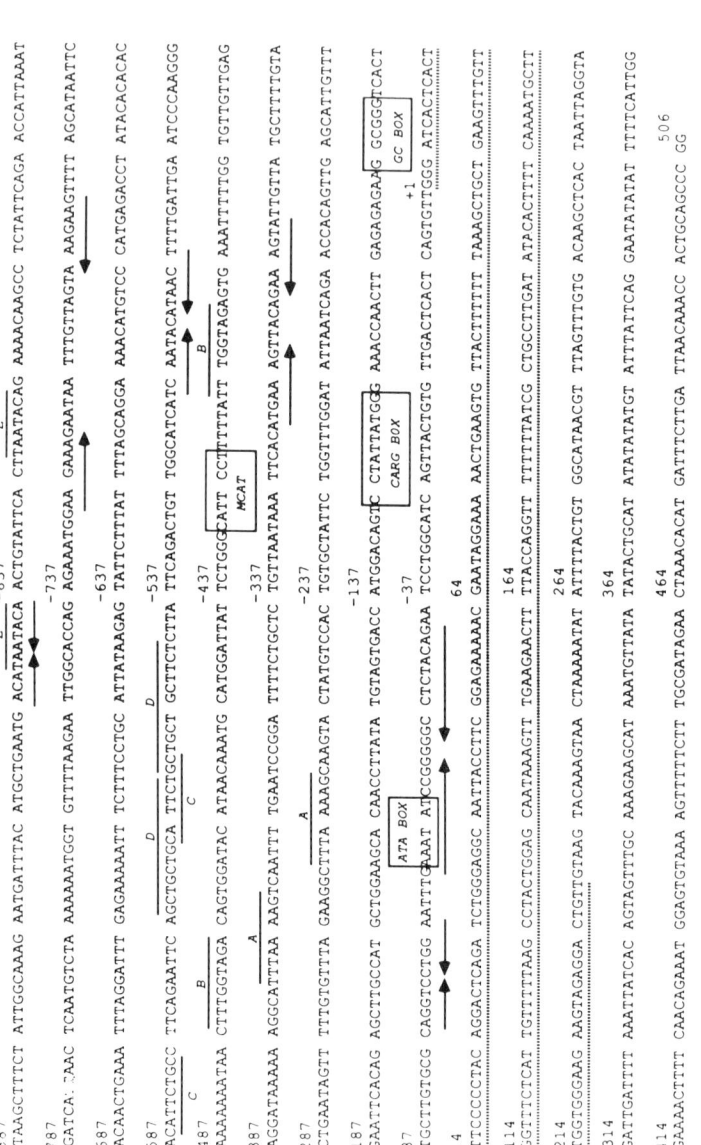

Fig. 4 The nucleotide sequence of the 1.4 kb *Hind* III-*Pst* 1 (HP2) subclone containing exon 1 described in Figure 3. Nucleotides corresponding to exon 1, determined by comparing the sequence of the HP2 fragment to cDNA sequences, are indicated by the underlying dotted line. Bases are numbered relative to the 5′ end of exon 1 (+1). The ATA box, GC box, CArG box and MCAT consensus sequences are boxed. Direct repeats (A to E) are overscored with solid lines. Repeat C (underlined) corresponds to the MCAT-like sequences described in the text. Inverted repeats are indicated by arrows.

the *DMD* gene can be predicted to lie in a region midway between the TATA box and the 5′ end of exon 1 derived from cDNA analysis, effectively extending the size of exon 1 by approximately 30 bp.

Pre-defined muscle-specific regulatory sequences located upstream of the basal promoter element include a CArG box at position −128 and a MCAT consensus at position −431 (Fig. 4). In addition, a potential CBAR element is found coincident with the CArG box (75% homology to the consensus sequence), and two identical MCAT-like sequences are located at positions −559 and −586 (repeat C in Fig. 4). A number of other direct and inverted repeats and dyad symmetries are distributed throughout the sequence, although their functional roles (if any) have not been defined. In addition, a comparison of sequences upstream of the *DMD* gene with other published muscle-specific gene promoter sequences has revealed a number of small, positionally conserved nucleotide homologies. Interestingly, these homologies tend to be clustered around the TATA box, in the CArG and GC box region, and in the region surrounding the two MCAT-like sequences.

In order to determine if this putative DMD promoter region is capable of regulating gene transcription in a muscle-specific manner, the 1.4 kb *Hind* III-*Pst* 1 (HP2) fragment was inserted in front of a promoterless *chloramphenicol acetyltransferase* (*CAT*) reporter gene in the pBLCAT 3 vector[43] to generate the pHP2CAT construct. The regulation of *CAT* gene expression was examined by introducing pHP2CAT into human skeletal muscle cell cultures using the modified calcium phosphate precipitation technique described by Chen and Okayama.[44] Cells were transfected at the myoblast stage and were allowed to differentiate into multinucleated myotubes prior to harvesting and assaying for CAT activity.[45] The results are shown in Figure 5. High levels of DMD promoter-mediated transient CAT expression were observed upon myoblast differentiation in all normal human primary and clonal muscle cell cultures studied (Fig. 5A, lanes 3 and 5 and others not shown). Normalization of CAT activity to cell extract protein concentrations in each assay (Fig. 5B) shows that a correlation exists between the degree of myoblast fusion and the transcriptional activity of the *DMD* gene promoter. Transcriptional activity in differentiated clonal human myotubes (lane 5) was 5-fold higher than that observed in predifferentiated clonal myoblasts (lane 4), and 2-fold higher than that observed in the differentiated primary muscle cell culture (in which the percentage

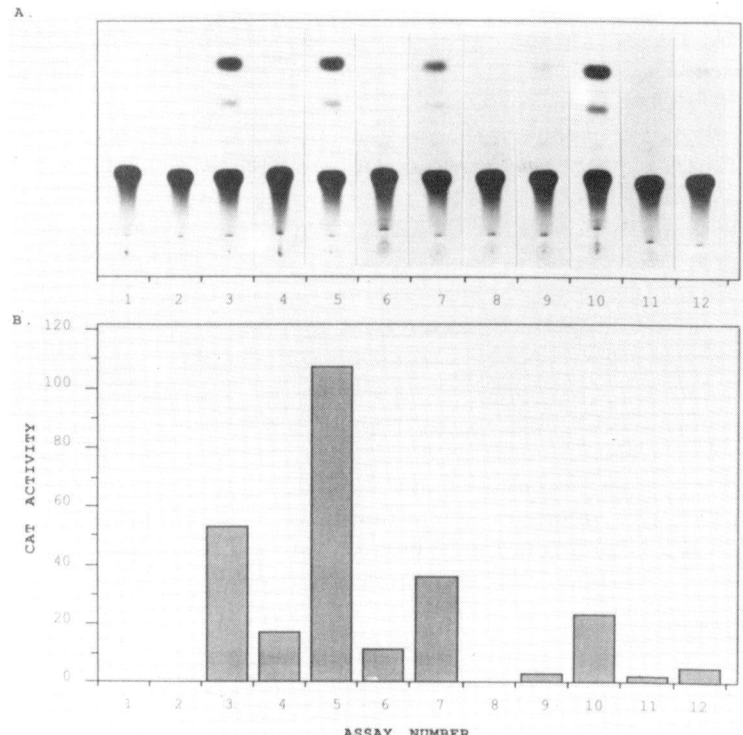

Fig. 5 Functional analysis of the *DMD* gene promoter region.
A. Transcriptional activity of the pHP2CAT construct in various human and rodent cell lines. Lanes: (1) control reaction containing no cell extract; (2) clonal human myotubes transfected with the promoterless pBLCAT 3 vector; (3) pHP2CAT in primary human myotubes; (4) pHP2CAT in clonal human myoblasts; (5) pHP2CAT in clonal human myotubes; (6) pHP2CAT in a poorly-differentiating clonal human myoblast line; (7) pSV2ACAT in clonal human myotubes; (8) pBLCAT3 in normal human fibroblasts; (9) pHP2CAT in normal human fibroblasts; (10) pHP2CAT in primary mouse myotubes; (11) pHP2CAT in C2 mouse myotubes; (12) pHP2CAT in L6 rat myotubes.
B. CAT assays depicted in Figure 5A expressed as the percent conversion of chloramphenicol to its acetylated forms/mg protein in cell extracts. Lane designations as in A.

of fused myotubes is lower due to the presence of non-myogenic cell types) (lane 3). Little or no CAT activity was observed in control reactions containing no cell extract (lane 1) or in clonal myotubes transfected with the promoterless pBLCAT3 vector (lane 2). Consistent with the above results, CAT activity measured in a non-fusing clonal human myoblast cell line (lane 6) was only 10% of that observed in the differentiating clonal myoblast culture (lane 5) grown and harvested under the same conditions. The relative strength of *DMD* gene promoter activity was examined by parallel transfections of clonal human myoblast cultures with a pSV2CAT construct[45] containing a functional SV40 promoter. CAT activity in pHP2CAT-transfected myotubes (lane 5) was found to be 3-fold higher than that of identical pSV2CAT-transfected cultures (lane 7). Transient expression of the pHP2CAT construct in normal human skin fibroblast cultures (lane 9) was only 4% of that observed in differentiated myotubes (lane 5), and only 19% of that observed in pre-differentiated myoblasts (lane 4). Although extremely low, DMD promoter activity in fibroblasts remained significantly higher than that of parallel cultures transfected with the promoterless pBLCAT3 plasmid (lane 8), suggesting that a basal level of *DMD* gene expression might occur in non-myogenic cells or that other regulatory elements are involved. The higher levels of CAT activity observed in pre-fusion myoblasts may be the result of a low level of differentiation in these cultures or may reflect an early stage of *DMD* gene activation. Together, these results indicate that the regulatory sequences which confer muscle-specific transcriptional activation to the *DMD* gene are located, at least in part, within these 887 bp upstream of the 5' end of exon 1.

The pHP2CAT construct was also introduced into neonatal mouse primary skeletal muscle cultures, as well as into the C2 mouse and L6 rat myoblast cell lines in order to determine whether functional regions within the *DMD* gene promoter have been conserved. CAT gene expression was clearly evident in differentiated mouse primary myoblast cultures transfected with pHP2CAT (Fig. 5, lane 10), indicating that sequences specifying differentiation-dependent expression of the *DMD* gene, and their associated *trans*-acting factors, have been conserved between humans and mice. The 2-fold lower CAT activity observed in primary mouse cultures as compared to that seen in primary human cultures (lane 3) may reflect the lower percentage of myoblast fusion attained by the primary mouse culture studied. In contrast to these results, no induction of DMD promoter-driven *CAT* gene expression was observed upon the differentiation of either the C2 mouse (lane 11) or the L6 rat (lane 12) myoblast cell lines, both of which display high levels of myotube formation. Low basal levels of DMD promoter activity in L6 rat myotubes is consistent with the absence of detectable levels of *DMD* mRNA in this cell line reported by Lev et al.[30] The L6 rat myoblast line is also incapable of inducing the human cardiac actin gene promoter upon differentiation,[3] but does support fusion-dependent activation of the rat embryonic myosin heavy chain gene promoter.[13] The C2 mouse myoblast cell line, on the other hand, has been used to demonstrate the activation of a number of muscle-specific gene promoters, including those of the human cardiac and skeletal actin genes,[3,4] the rat embryonic myosin heavy chain gene,[13] and the hamster desmin gene.[14] The inability of this cell line to induce *DMD* gene promoter activity upon differentiation is therefore somewhat surprising, especially in view of the presence of the CArG and MCAT consensus sequences. Since muscle-specific promoter function is dependent on the expression of specific *trans*-acting factors required for promoter function, these results indicate that neither the L6 nor the C2 myoblast cell lines express the appropriate *trans*-acting factor(s) required for *DMD* gene promoter induction. Appropriate regulation of *DMD* gene expression must therefore require additional factors which recognize as yet undefined *cis*-acting promoter elements upstream of the *DMD* gene, or may involve factors having altered affinities for defined promoter elements such as the CArG box found in many of the major contractile and cytoskeletal protein gene promoters studied to date.

FUTURE PROSPECTS

The successful cloning and characterization of genomic and cDNA sequences corresponding to the *DMD* gene, and the subsequent identification of dystrophin as the protein product of this gene has provided the means by which to begin to approach an understanding of the pathogenesis of this disease. Some insights into the functional role of dystrophin have already been provided from an analysis of its amino acid sequence, which predicts a long rod-shaped protein consisting of four domains defined as an amino-terminal actin-binding domain,[46] followed by a large spectrin-like repeat domain,[47] a cysteine-rich region containing two potential calcium binding sites, and a unique carboxy-terminal domain which does not exhibit homology to any known protein sequences.[19] Taken together with the immunocytochemical localization of dystrophin to the sarcolemma and possibly the t-tubular system of skeletal muscle,[35-38] these results suggest that this protein is associated with the muscle membrane cytoskeletal network and may be important in the management of contraction-induced stress.

The demonstration that *DMD* gene transcription is initiated with myoblast differentiation in vitro indicates that myogenic cultures can be used as an experimental system within which to approach the precise definition of the functional role of this protein and the mechanisms which underly its transcriptional regulation. In this report we have presented our initial observations of dystrophin expression and localization in clonal human myogenic cultures. Our observation of a patchy and longitudinal orientation of dystrophin and its preferential association with the surface membrane in developing myotubes may represent the early stages of assembly of dystrophin into the membrane cytoskeletal network. A variety of approaches can now be used to investigate this possibility further. Dual immunofluorescence labelling, immunoelectron microscopy and immunoprecipitation techniques are particularly applicable to studies of dystrophin interactions in normal and *DMD* patient cultures. As additional antibodies prepared to various regions of the dystrophin molecule are available, a more precise orientation and definition of potential interactive regions at the cell surface may be possible. Furthermore, immunohistochemical analysis of myogenic cultures derived from patients having defined deletions in the *DMD* gene may prove useful in the identification of important functional domains within the dystrophin molecule.

Any future application of gene transfer techniques to the treatment of DMD will require a complete understanding of the tissue-specific transcriptional regulation of this gene. The introduction of DMD promoter-driven reporter gene constructs into transgenic animals will facilitate initial evaluations of the potential of this therapeutic course. Our finding that CArG and MCAT regulatory sequences are present in the promoter region of the *DMD* gene, and that this region is capable of directing muscle-specific gene expression strongly implies, but does not prove, that these sequences are involved in the myogenic regulation of *DMD* gene transcription. Deletion and mutational analyses, along with factor binding assays will more precisely define the transcriptional role of these elements and will aid in the identification of other potential regulatory elements required for the myogenic regulation of this gene. In addition, the mechanisms underlying *DMD* gene transcriptional regulation in cardiac, smooth muscle and nerve cells must also be considered. Cell lines and culture techniques which support the growth of these cells as primary cultures in vitro are readily available and can be applied in similar studies toward the elucidation of regulatory elements required for expression in these cell types.

An alternative to gene replacement therapy for the treatment of DMD has been suggested by the recent demonstration by Law et al. that transplantation of normal myoblasts into dystrophic muscle results in some alleviation of muscle weakness in murine dystrophy.[48] As the effectiveness of this treatment depends on the relative contribution of normal gene transcripts from transplanted nuclei, there is considerable merit in experiments designed to maximize expression of the normal gene. Possible strategies can be envisioned for the establishment of transplantable myogenic lines in which dystrophin gene transcription is amplified through the introduction of specific regulatory enhancer elements upstream of the *DMD* gene either directly by homologous recombination or indirectly through stable integration of a functional mini-gene construct. Whether or not myoblast transplantation becomes an effective treatment for Duchenne and Becker muscular dystrophy awaits further animal experimentation and carefully controlled clinical trials.

REFERENCES

1 Dynan WS, Tjian R. Control of eukaryotic messenger RNA synthesis by sequence-specific DNA binding proteins. Nature 1985; 316: 774–778
2 Ptashne M. Gene regulation by proteins acting nearby and at a distance. Nature 1986; 322: 697–701
3 Miwa T, Boxer LM, Kedes L. CArG boxes in the human cardiac alpha-actin gene are core binding sites for positive trans-acting regulatory factors. Proc Natl Acad Sci USA 1987; 84: 6702–6706
4 Muscat GEO, Kedes L. Multiple 5'-flanking regions of the human alpha-skeletal actin gene synergistically modulate muscle-specific expression. Mol Cell Biol 1987; 7: 4089–4099
5 Arnold HH, Tannich E, Paterson BM. The promoter of the chicken cardiac myosin light chain 2 gene shows cell-specific expression in transfected primary cultures of chicken muscle. Nucleic Acids Res 1988; 16: 2411–2429
6 Mar JH, Antin PB, Cooper TA, Ordahl CP. Analysis of the upstream regions governing expression of the chicken cardiac troponin T gene in embryonic cardiac and skeletal muscle cells. J Cell Biol 1988; 107: 573–585
7 Sternberg EA, Spizz G, Perry WM, et al. Identification of upstream and intragenic regulatory elements that confer cell-type-restricted and differentiation-specific expression on the muscle creatine kinase gene. Mol Cell Biol 1988; 8: 2896–2909
8 Gustafson TA, Miwa T, Boxer LM, Kedes L. Interaction of nuclear proteins with muscle-specific regulatory sequences of the human cardiac alpha-actin promoter. Mol Cell Biol 1988; 8: 4110–4119
9 Muscat GEO, Gustafson TA, Kedes L. A common factor regulates skeletal and cardiac alpha-actin gene transcription in muscle. Mol Cell Biol 1988; 8: 4120–4133
10 Grichnik JM, French BA, Schwartz RJ. The chicken skeletal alpha-actin gene

promoter region exhibits partial dyad symmetry and a capacity to drive bidirectional transcription. Mol Cell Biol 1988; 8: 4587–4597

11 Carroll Sl, Bergsma DJ, Schwartz RJ. A 29-nucleotide DNA segment containing an evolutionarily conserved motif is required in cis for cell-type-restricted repression of the chicken alpha-smooth muscle actin gene core promoter. Mol Cell Biol 1988; 8: 241–250

12 Mar JH, Ordahl CP. A conserved CATTCCT motif is required for skeletal muscle-specific activity of the cardiac troponin T gene promoter. Proc Natl Acad Sci USA 1988; 85: 6404–6408

13 Bouvagnet PF, Strehler EE, White GE, et al. Multiple positive and negative 5' regulatory elements control the cell-type-specific expression of the embryonic skeletal myosin heavy-chain gene. Mol Cell Biol 1987; 7: 4377–4389

14 Piper FR, Slobbe RL, Ramaekers FCS, et al. Upstream regions of the hamster desmin and vimentin genes regulate expression during in vitro myogenesis. EMBO J 1987; 6: 3611–3618

15 Kunkel LM, Monaco AP, Middlesworth W, et al. Specific cloning of DNA fragments absent from the DNA of a male patient with an X chromosome deletion. Proc Natl Acad Sci USA 1985; 82: 4778–4782

16 Ray PN, Belfall B, Duff C, et al. Cloning of the breakpoint of an X:21 translocation associated with Duchenne muscular dystrophy. Nature 1985; 318: 672–675

17 Monaco AP, Neve RL, Colletti-Feener C, et al. Isolation of candidate cDNAs for portions of the Duchenne muscular dystrophy gene. Nature 1986; 323: 646–650

18 Burghes AHM, Logan C, Hu X, et al. A cDNA clone from the Duchenne/-Becker muscular dystrophy gene. Nature 1987; 328: 434–437

19 Koenig M, Monaco AP, Kunkel LM. The complete sequence of dystrophin predicts a rod-shaped cytoskeletal protein. Cell 1988; 53: 219–228

20 Gardner-Medwin D. Clinical features and classification of the muscular dystrophies. Br Med Bull 1980; 36: 109–115

21 Roses AD, Appel SH. Inherited membrane disorders of muscle: Duchenne muscular dystrophy and myotonic muscular dystrophy. In: Andreoli TE et al., eds. Physiology of membrane disorders. New York: Plenum Medical, 1978: pp. 801–815

22 Emery AEH. Duchenne muscular dystrophy. Oxford: Oxford University Press, 1979

23 Brown CS, Thomas NST, Sarfarazi M, et al. Genetic linkage relationships of seven DNA probes with Duchenne and Becker muscular dystrophy. Hum Genet 1985; 71: 62–74

24 Koenig M, Hoffman EP, Bertelson CJ, et al. Complete cloning of the Duchenne muscular dystrophy (DMD) cDNA and preliminary genomic organization of the DMD gene in normal and affected individuals. Cell 1987; 50: 509–517

25 Dunnen JT den, Bakker E, Klein Breteler EG, et al. Direct detection of more than 50% of the Duchenne muscular dystrophy mutations by field inversion gels. Nature 1987; 329: 640–642

26 Oronzi Scott M, Sylvester JE, Heiman-Patterson T, et al. Duchenne muscular dystrophy gene expression in normal and diseased human muscle. Science 1988; 239: 1418–1420

27 Nudel U, Robzyk K, Yaffe D. Expression of the putative Duchenne muscular dystrophy gene in differentiated myogenic cell cultures and in the brain. Nature 1988; 331: 635–638

28 Chamberlain JS, Pearlman JA, Muzny DM, et al. Expression of the murine Duchenne muscular dystrophy gene in muscle and brain. Science 1988; 239: 1516–1417

29 Chelly J, Kaplan JC, Maire P, et al. Transcription of the dystrophin gene in human muscle and non-muscle tissues. Nature 1988; 333: 858–860

30 Lev AA, Feener CC, Kunkel LM, Brown RH. Expression of the Duchenne's muscular dystrophy gene in cultured muscle cells. J Biol Chem 1987; 262: 14817–15820

31 Hoffman EP, Brown RH, Kunkel LM. Dystrophin: The protein product of the Duchenne muscular dystrophy locus. Cell 1987; 51: 919–928

32 Hoffman EP, Hudecki MS, Rosenberg PA, et al. Cell and fiber type distribution of dystrophin. Neuron 1988; 1: 411–420

33 Hoffman EP, Knudson CM, Campbell KP, Kunkel LM. Subcellular fractionation of dystrophin to the triads of skeletal muscle. Nature 1987; 330: 754–758

34 Knudson CM, Hoffman EP, Kahl SD, et al. Evidence for the association of dystrophin with the transverse tubular system in skeletal muscle. J Biol Chem 1988; 263: 8480–8484

35 Zubrzycka-Gaarn EE, Bulman DE, Karpati G, et al. The Duchenne muscular dystrophy gene product is localized in sarcolemma of human skeletal muscle. Nature 1988; 333: 466–469

36 Arahata K, Ishiura S, Ishiguro T, et al. Immunostaining of skeletal and cardiac muscle cell surface membrane with antibody against Duchenne muscular dystrophy peptide. Nature 1988; 333: 861–863

37 Bonilla E, Samitt CE, Miranda AF, et al. Duchenne muscular dystrophy: Deficiency of dystrophin at the muscle cell surface. Cell 1988; 54: 447–452

38 Watkins SC, Hoffman EP, Slayter HS, Kunkel LM. Immunoelectron microscopic localization of dystrophin in myofibres. Nature 1988; 333: 863–866

39 Blau HM, Webster C. Isolation and characterization of human muscle cells. Proc Natl Acad Sci USA 1981; 78: 5623–5627

40 Blau HM, Webster C, Pavlath GK. Defective myoblasts identified in Duchenne muscular dystrophy. Proc Natl Acad Sci USA 1983; 80: 4856–4860

41 Tapscott SJ, Davis RL, Thayer MJ, et al. Myo DI: A nuclear phosphoprotein requiring a Myc homology region to convert fibroblasts to myoblasts. Science 1988; 242: 405–411

42 Smith TJ, Forrest SM, Cross GS, Davis KE. Duchenne and Becker muscular dystrophy mutations: analysis using 2.6 kb of muscle cDNA from the 5' end of the gene. Nucleic Acids Res 1987; 15: 9761–9769

43 Luckow B, Schutz G. CAT constructions with multiple unique restriction sites for the functional analysis of eukaryotic promoters and regulatory elements. Nucleic Acids Res 1987; 15: 5490

44 Chen C, Okayama H. High-efficiency transformation of mammalian cells by plasmid DNA. Mol Cell Biol 1987; 7: 2745–2752

45 Gorman CM, Moffat LF, Howard BH. Recombinant genomes which express chloramphenicol acetyltransferase in mammalian cells. Mol Cell Biol 1982; 2: 1044–1051

46 Hammonds RG. Protein sequence of DMD gene is related to actin-binding domain of alpha-actinin. Cell 1987; 51: 1

47 Davison MD, Critchley DR. Alpha-actinins and the DMD protein contain spectrin-like repeats. Cell 1988; 52: 159–160

48 Law PK, Goodwin TG, Wang MG. Normal myoblast injections provide genetic treatment for murine dystrophy. Muscle Nerve 1988; 11: 525–533

British Medical Bulletin (1989) Vol. 45, No. 3, pp. 703–718
© The British Council 1989

Animal models of Duchenne and Becker muscular dystrophy

B J Cooper
Department of Pathology, New York State College of Veterinary Medicine, Cornell University, Ithaca, NY, USA

Two animal models have been shown to be related to Duchenne and Becker muscular dystrophy at the molecular level. The *mdx* mouse is characterized by early onset of muscle degeneration and very mild clinical disease. The disease is minimally progressive and fibrosis of muscle is absent. Linkage studies, absence of dystrophin, and reduced levels of message indicate that the mutation in *mdx* lies in the gene for dystrophin, the gene that is defective in Duchenne and Becker muscular dystrophy. The *xmd* dog develops lesions that are essentially indistinguishable from those of Duchenne dystrophy, and there is progressive fibrosis and destruction of muscle tissue. Affected dogs develop severe clinical disease. The absence of dystrophin and its message in muscle, and the linkage of RFLPs recognized by Duchenne cDNA probes, indicate that the mutation in the *xmd* dog lies in the gene for dystrophin. Exploitation of these models should lead to a greater understanding of molecular and cellular events involved in the pathogenesis of Duchenne and Becker muscular dystrophies.

Animal models have contributed significantly to the investigation and understanding of a number of biological and pathobiological processes. The potential advantages of an animal model of Duchenne-type muscular dystrophy have led, over the years, to the investigation of several models. These include the dystrophic mouse, the dystrophic hamster, and the dystrophic chicken. These models, which have been reviewed previously,[1,2] have contributed

0007–1420/89/0045–0703/$10.00

useful information on the pathobiology of muscle but for various reasons are not ideal models for Duchenne muscular dystrophy (DMD). In particular, each has significant phenotypic dissimilarities to DMD and each is inherited as an autosomal recessive trait, unlike DMD which is inherited as an X-linked recessive. The features of an 'ideal' model of DMD, in which questions relevant to the pathogenesis and therapy of the human disease can be addressed, would include similar clinical signs and progression, similar lesions, and a similar molecular basis. Because of the strong conservation of genetic information on the X chromosome,[3] it is unlikely that any model not inherited as an X-linked trait will have a similar molecular basis to DMD. For this reason this review will be limited to a discussion of two recently discovered X-linked models, the *mdx* mouse, and the *xmd* dog.

THE *MDX* MOUSE

The mouse mutant *mdx* was discovered by Bulfield and colleagues in a colony of C57Bl/10ScSn inbred mice which was being screened for the enzyme pyruvate kinase.[4] Animals were found which had elevated blood levels of the enzyme. In breeding experiments this trait segregated as an X-linked recessive mutation. Enzyme elevations were recognized to be due to the muscle-type isoenzyme and were associated with necrosis of muscle tissue. More recently, two additional *mdx* mutations were identified in a group of mice treated with ethylnitrosourea (ENU) and monitored for enzyme elevations (Caskey, personal communication, see Ref. 5).

Clinical signs

Mice affected by the *mdx* mutation show relatively few signs of clinical disability. Bulfield noted that a 12 month old mutant mouse showed muscular tremors and mild incoordination.[4] Torres and Duchen observed mild weakness, evidenced by reduced ability of affected animals to cling to the bars of a wire grid.[6] Other investigators have stated that affected *mdx* mice are essentially normal.[7-9] In fact, Carnwath and Shotton[9] subjected *mdx* mice to 1.5 miles per day of voluntary running without apparent exacerbation of the disease. Similarly, lifespan and reproductive performance are minimally affected by the *mdx* mutation. Some authors have reported no diminution in breeding capacity or longevity,[9] while others have seen slight reductions in litter size and increased neonatal mortality.[6] Despite these slight inconsistencies, it is clear from the literature that mice affected by the *mdx* mutation show little if any deleterious effects.

Pathology

A number of publications which report the nature and progression of lesions in the *mdx* mouse have now appeared.[4,6–11] The most striking lesions occur in skeletal muscle tissue. These consist of necrosis of muscle fibres followed by infiltration and phagocytosis by macrophages (Fig. 1) and regeneration of muscle fibres (Fig. 2). The majority of investigators have found that the onset of degenerative lesions begins at approximately 20 days of age. Dangain and Vrbova found that *mdx* mice suffered massive, abrupt muscle fibre necrosis beginning at about 3 weeks of age,[7] followed by rapid regeneration leading to complete recovery by 5 weeks of age. Although details vary, most investigators agree that there is early degeneration and necrosis of muscle fibres, accompanied by infiltration by macrophages, and followed by efficient regeneration. Torres and Duchen,[6] in a very detailed study of neuromuscular lesions in the *mdx* mouse, found large, eosinophilic fibres and ultrastructural disorganization of the Z line at 1 day of age, and recognizable necrosis and histiocytic infiltration by 5 days of age. Necrosis reached a peak at 5–6 weeks of age, but was present to a lesser degree throughout the life of the animal. Regeneration was

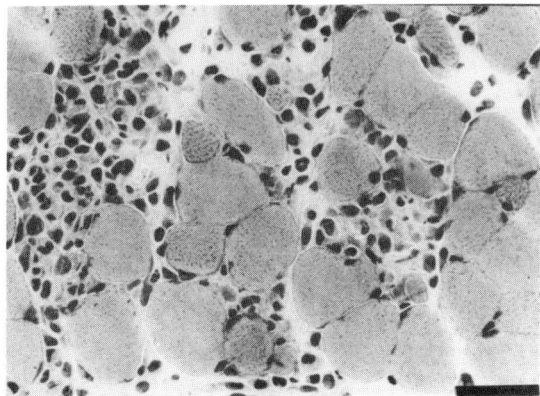

Fig. 1 Gastrocnemius muscle of *mdx* mouse at 18 days of age; necrotic fibres are infiltrated by macrophages. Frozen section, hematoxylin and eosin, calibration bar = 50 μm. (Reproduced, with permission, from Karpati et al.[11]).

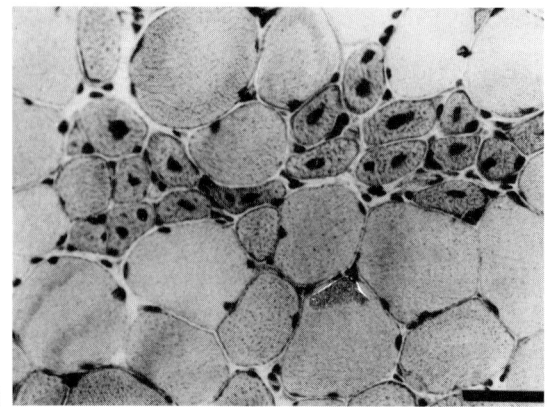

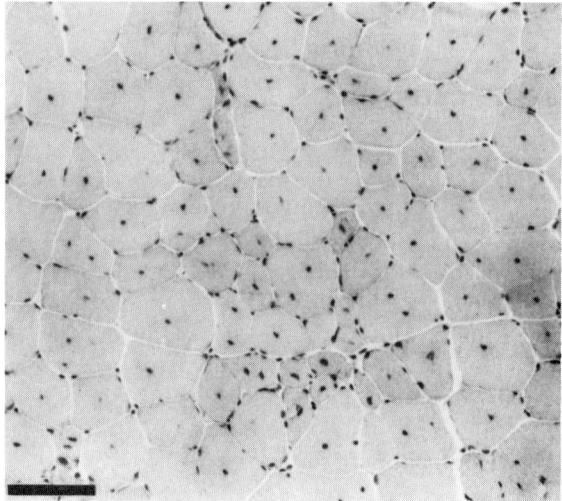

Fig. 2 Gastrocnemius muscle of *mdx* mouse: (a) 45 days of age; a group of regenerating fibres characterized by small caliber, central nuclei, and cytoplasmic basophilia; (b) 60 days of age—approximately 80% of fibres are centronucleated. Frozen section, hematoxylin and eosin, calibration bar = 50 μm. (Reproduced, with permission, from Karpati et l.[11]).

recognized by 10 days of age and was present at all ages studied. Most investigators have reported that the most dramatic lesions in the *mdx* mouse occur at about 20 days of age, and that there is ongoing, but less severe, degeneration and regeneration throughout the life of the animal. These lesions are accompanied by elevations in pyruvate kinase (PK) and creatine kinase (CK) which also peak at about 5 weeks of age, but remain elevated throughout the life of the animal.[12] All reports describing lesions in this mutant agree that one of the cardinal features of advanced DMD, namely the presence of fibrosis and fatty infiltration, is essentially absent.[6,8,10,13]

A striking feature of muscle lesions in the *mdx* mouse is the persistence of centrally located nuclei in regenerated muscle fibres. This change is generally accepted to indicate previous degeneration and regeneration. In the *mdx* mouse, the number of centrally nucleated muscle fibres increases with age until the majority of fibres have this appearance.[6,8,9,11,13,14] To several investigators this suggests that regeneration is more efficient in this mutant than in patients with DMD, but the progressive decline in the number of necrotic fibres, and in serum CK levels, suggests that the best interpretation is that regenerated muscle fibres in the *mdx* mouse are relatively stable and less susceptible to injury than the original fibres they replace. Indeed, it has been shown by autoradiography after administration of tritiated thymidine that the number of actively regenerating cells is greatly reduced in mature (32 weeks old) *mdx* mice compared to young (4 week old) animals. Thus, the evidence suggests that both degenerative and regenerative activity decline in concert as affected mice mature.

Although the *mdx* mutant has minimal clinical signs, some functional alterations in the affected muscles have been reported. Dangain and Vrbova reported reduced tension output and prolonged half relaxation time in mutant mice 3–4 weeks of age.[7] In contrast, there was no difference between adult *mdx* and controls. Coulton et al.[15] also found a reduction in isometric force produced by muscle of young *mdx* mice compared to controls when measurements were corrected for cross-sectional areas of muscle. As regenerated muscle fibres in older mice still bear the genetic defect, it is likely that these differences in muscle strength are due to loss of functional fibres within the muscle due to necrosis.

Cardiac lesions, which are found in at least some patients with DMD, have been reported to be absent in the *mdx* mouse by some authors,[6,8] while others have reported degeneration and necrosis of cardiac myocytes.[10,16]

In summary, the essential morphologic features shared by the *mdx* mouse and human patients with DMD are the presence of early onset degeneration and necrosis of muscle, followed by infiltration by macrophages, and regeneration. The disease in the mouse, however, is clinically and pathologically much milder and less progressive than that in man. The fibrosis and fatty infiltration of muscle which is so prominent in muscle of older DMD patients is absent in the mouse. In the *mdx* mouse, regenerated muscle retains its central nucleation over a long period, and appears to be more stable than that in man. The inconsistencies in the literature regarding the exact age of onset of lesions in the mouse probably can be attributed to variations in sampling techniques. The disagreement over the presence or absence of cardiac lesions is more difficult to explain, and is a point which needs to be clarified.

Molecular pathology

The identification of the gene which is defective in DMD,[17,18] its gene product, dystrophin,[19] and the development of molecular probes which recognize the gene, its message, and its protein product have provided an almost unprecedented opportunity to study the molecular relationship between the human disease and putative animal models.

The gene for dystrophin has been shown to be highly conserved in the mouse, the nucleic acid sequences being about 90% homologous.[20] Early attempts to determine the relationship between *mdx* and DMD relied on classical chromosomal mapping techniques. Initial studies demonstrated linkage between the *mdx* gene and markers which are known to be located on the long arm of the human X chromosome, near the mutation that is responsible for Emery-Dreifuss muscular dystrophy (EMD). This led to the suggestion that the *mdx* myopathy might be homologous to the milder *EMD*, rather than *DMD*.[21] However, further linkage studies using interspecific crosses and a number of X chromosomal markers have demonstrated that the organization of genes on the murine X chromosome is somewhat rearranged compared to the human, and that the murine homologue of the *DMD* locus (*mDMD*) maps close to the *mdx* mutation.[22–24] In a more recent linkage study, again using interspecific crosses and a variety of X chromosomal markers, recombination was obtained within the *mDMD* locus.[25] The recombinational breakpoints were on either side of the *mdx* mutation, providing the first direct evidence that the murine disease is due to a mutation in the dystrophin gene.

Further evidence for homology between the *mdx* gene and the *DMD* gene is provided by the fact that mRNA for dystrophin is expressed at much lower than normal levels in muscle and brain of mutant mice. Using the RNase protection technique, Chamberlain et al.[5] found mRNA at approximately 20% of normal levels. On northern blots the message, although reduced in amount, was shown to be of normal size. The protein product of the *DMD* gene, dystrophin, has also been shown to be absent from muscle of the *mdx* mutant, both by western blotting,[19] and by immunostaining.[26,27] Together, these data indicate that the myopathy in the *mdx* mouse is due to a defect in the *DMD* locus leading to the absence of the membrane-associated muscle protein, dystrophin.

The exact nature of the *mdx* mutation is as yet unknown. The lack of abnormalities on Southern analysis of the mutant *mDMD* gene[22,25] suggests that no gross alterations of the gene, such as large deletions, are present, and it is likely that the murine disease is due to a point mutation. The recent approximate localization of the *mdx* mutation within the *mDMD* gene should lead to characterization of the mutation. Whether the 'new' *mdx* strains will prove to have similar mutations remains to be seen.

THE *XMD* DOG

In 1983 we studied a young male Golden Retriever dog which presented with clinical evidence of myopathy.[28] This animal led to the recognition of the canine model of muscular dystrophy. We have called the disease Canine X-linked Muscular Dystrophy (CXMD), and the model the *xmd* dog.

This disease has been known for a number of years. The first probable reported case was a 10-day-old Golden Retriever pup included in a review of myopathies in animals.[29] Since then several other cases have been reported.[28,30–32] Including those made known to us by colleagues, about 12 cases have occurred naturally. All of these have been male, suggesting X-linked inheritance. Because of the clinical and pathologic similarities between the canine disease and DMD, and the known conservation of genes on the X chromosome, we postulated that the two diseases might be related. We therefore decided to establish a colony of dogs bearing the trait in order to characterize the disease, and establish its relationship to DMD.

Inheritance

The breeding colony was established by mating an affected male to normal unrelated female dogs. Putative carrier females from the F1 generation were retained, and bred back to the original sire. Affected progeny were obtained in this litter. To definitively test the supposed X-linked pattern of inheritance, two proven carrier bitches were bred to unrelated normal male dogs. From these breedings 4 of 7 males were affected, while all 10 females were normal. This pattern supported an X-linked mode of inheritance.[33] Cytogenetic studies revealed no chromosomal abnormalities, the X chromosomes, in particular, being normal.

Clinical signs

Detailed clinical studies on affected dogs have been carried out in our laboratories.[34] The results are generally consistent with those reported by other investigators.[30,31] Clinically evident abnormalities first appear at approximately 8 weeks of age. Affected pups are moderately weak, have limited opening of the jaw, and develop a stiff gait. Before 8 weeks it is difficult to detect definitive signs of the disease, but, with experience, affected pups can be tentatively identified based on reluctance to move around, and on their small stature compared to littermates. Impaired growth in affected pups has been documented.[34] Clinical signs progress over several months to include distinct muscle atrophy, abduction of the elbows, adduction of the hocks, overextension of the carpi, and overflexion of the tarsi. Older dogs have reduced range of motion in joints, associated with muscle contracture. Affected dogs have a weak bark, tire easily on exercise, and have an increased resting respiratory rate with obvious abdominal breathing. Older affected dogs frequently develope distortion of the spinal column and thoracic wall.

Clinically, the disease tends to stabilize at about 6 months of age. There is some variation in the severity of clinical signs in the *xmd* dog, in particular between affected dogs in different litters. In addition, homozygous affected females tend to show somewhat milder clinical signs than male littermates.[34]

Affected dogs have consistently, and often dramatically, elevated serum creatine kinase (CK) levels. Aspartate aminotransferase (AST) and alanine aminotransferase (ALT) may also be elevated (Valentine, unpublished observations). Elevations of CK can be detected by 2 days of age, indicating that muscle lesions are present at, or very soon after, birth.[34] In fact, occasional pups develop a clinically fulminant form of the disease and die within a few days of birth. These pups have severe lesions, described below. Serum CK levels remain elevated throughout the life of the animal, although they progressively decline in the later stages of the disease. Although very few animals have been studied over such a prolonged period, some dystrophic dogs can live as long as 5–6 years.

In *xmd* dogs a moderate amount of exercise produces dramatic elevations of CK above the already high resting levels (Valentine, unpublished observations). This suggests that, at least in the dog, muscle lacking dystrophin is abnormally sensitive to exercise induced injury. Whether this is due to inherent fragility of the muscle cell membrane or some other mechanism remains to be determined.

Dogs with CXMD frequently develop cardiomyopathy. This is apparent as reduced myocardial contractility on echocardiographic examination.[34] Older dogs may develop clinical signs of heart failure. For example, the male from which our colony was established died of congestive heart failure at 6 years of age.

Pathology

Lesions in the *xmd* dog are most prominent in skeletal muscles, and are strikingly similar to those of DMD.[28,31,35] They consist of muscle degeneration and necrosis, characterized in early stages of development by swollen, hyalinized fibres. These changes progress to necrosis, infiltration and phagocytosis by macrophages, and extensive muscle fibre regeneration (Fig. 3). Lesions tend to develop earlier and to be most severe in certain muscle groups, including the trapezius, deltoideus, brachiocephalicus, extensor carpi radialis, and sartorius muscles, and the diaphragm. These muscles are also severely affected in pups with severe neonatal disease.[34] Muscle from dogs with more advanced disease show a wide range of fibre sizes, from abnormally small to hypertrophic.[31] In contrast to the *mdx* mouse, the dog has a progressive disease which results in eventual fibrosis of muscle tissue (Fig. 4). In fact, it appears to be these changes in muscles of affected dogs which results in the 'stiff' gait which characterizes the disease clinically. In the dog, regenerating muscle fibres are characteristically small, with basophilic cytoplasm and large, vesicular nuclei. However, central nucleation is relatively transient, and nuclei appear to adopt a peripheral location before the regenerating fibre reaches normal size (Fig. 3).

Lesions similar to so-called 'delta lesions' in DMD patients[36,37] also occur in the dog. Ultrastructurally, these are characterized by subplasmalemmal lysis of cytoplasmic components, sometimes associated with discontinuities of the plasma membrane.[28] These

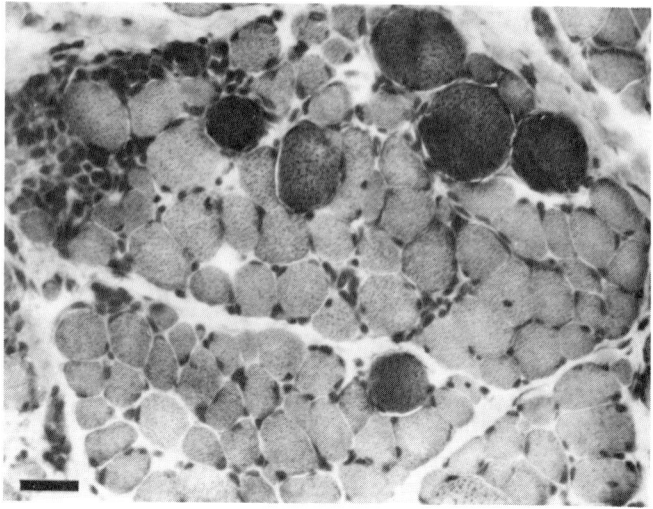

Fig. 3 Biceps femoris muscle, 17 week *xmd* dog. There are several acutely injured, swollen, darkly stained fibres, and several necrotic fibres infiltrated by macrophages. Note the marked variation in fibre size. Trichrome stain, calibration bar = 10 μm

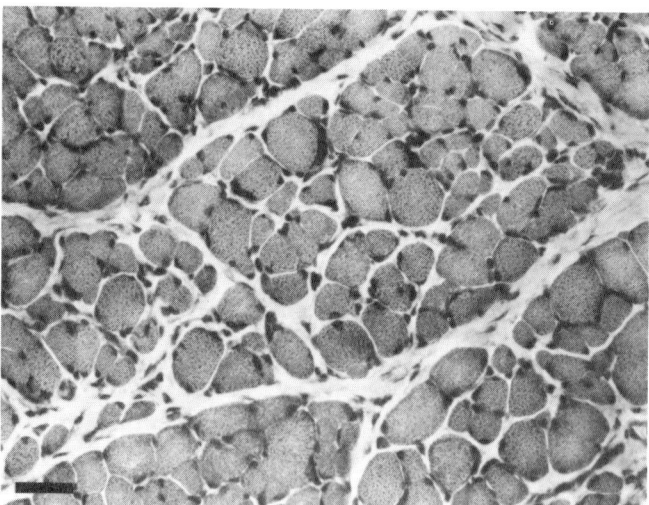

Fig. 4 Sartorius muscle, 30 day-old *xmd* dog. There is already obvious interstitial fibrosis. The sartorius muscle is affected early in the disease process. Trichrome stain, calibration bar = 10 μm

are often associated with hypercontracted myofibrils, presumably corresponding to the swollen, hyalinized fibres seen at the light microscopic level. As in muscle from DMD patients,[36] acutely injured fibres usually stain positively for calcium, often demonstrating a wedge-like subplasmalemmal distribution.[38]

As indicated above, cardiomyopathy appears to occur consistently in the *xmd* dog. We have had the opportunity to study hearts from a number of older dystrophic dogs and have found lesions similar to those reported in DMD patients.[39,40] In these dogs there was marked fibrosis of the subepicardial myocardium (unpublished observations). This lesion is most pronounced around interfascicular vascular bundles, but dissection of individual muscle cells by fibrous tissue can be seen. Recognizable necrosis of cardiac myocytes is rare. However, we have seen some dogs with severe myocardial necrosis apparently resulting from stressful situations.

Molecular pathology

The striking phenotypic similarities between CXMD and DMD and their common X-linked mode of inheritance, suggested that the two diseases might be due to a defect in the same gene. This hypothesis is supported by the data presently available.[35] We have shown that cDNA probes[41] for the human dystrophin gene recognize a 14 kb transcript in muscle tissue of normal dogs similar to that of man. Dogs with CXMD lack this transcript. In addition, analysis by western blotting has shown that dystrophin is undetectable in *xmd* muscle. Finally, we have identified RFLPs recognized by human dystrophin cDNA probes that segregate with the disease phenotype. Thus, although the mutation in the *xmd* dog has not yet been defined, there is compelling evidence that the disease is due to a defect in the dystrophin gene. These results are supported by those of immunocytochemical studies in which we have confirmed that immunoreactive dystrophin cannot be detected in tissue sections of dystrophic canine muscle.

LESSONS FROM THE MODELS

Because both of these models were discovered relatively recently, much of the effort to date has gone into characterization, at both the phenotypic and molecular levels. Each will play an important role in studying the pathogenesis of the human disease, and in approaches to its therapy.

Some recent studies have utilized the *mdx* mouse to address certain aspects of the pathogenesis of muscular dystrophy of the Duchenne type. Studies by Karpati have confirmed that muscle fibres of small caliber are relatively resistent to degeneration.[11] Although the reasons for this are unclear it seems possible that the mechanical stresses placed on the plasma membrane are reduced in small fibres. Freeze fracture studies of the plasma membrane of *mdx* mice have revealed decreased density of orthogonal arrays, disassociation of surface caveolae from the I band, and variable reduction in intramembranous particles (IMPs).[42] These changes are very similar to those reported in Duchenne dystrophy.[43–46] However, their significance in relation to the absence of dystrophin is still difficult to judge. In particular, the youngest mice studied were 8 weeks old, which is well after the peak occurrence of lesions. Thus it is still difficult to dissociate primary changes in the dystrophic sarcolemma with those due to degeneration and regeneration. Activity of proteolytic enzymes in muscle of *mdx* mice have also been studied. Sawada et al.[47] found early elevations of a muscle-specific trypsin-like enzyme in *mdx* muscle, and Sano et al. have immunocytochemically localized cathepsins B, H, and L in muscle of this mutant.[48] However, administration of the cathepsin B inhibitor, chymostatin, had no apparent beneficial effect on *mdx* mice.[49]

Some particularly significant results concerning the role of membrane dysfunction and regulation of intracellular calcium have very recently been obtained in the *mdx* mouse. Using the fluorescent calcium chelator fura-2, Turner et al.[50] found that intracellular calcium concentrations were elevated in *mdx* muscle fibres, and that the kinetics of calcium transients in response to stimulation were slowed. Thus, during stimulation of *mdx* muscle fibres, intracellular calcium levels remained elevated longer, and basal levels were higher, than in control mice. Furthermore, these investigators showed that there was an increase in rate of degradation of muscle protein in *mdx* mice which could be correlated with intracellular calcium concentrations. Presumably this process is mediated through calcium-activated proteases.[50] These results are consistent with long-held hypotheses that there is a plasmalemmal defect in dystrophic muscle fibres,[37,51–53] which we can now suggest result directly or indirectly from the absence of dystrophin.

In the *xmd* dog we have been studying the expression of dystrophin in muscle from carrier females. These cells would be expected to be natural mosaics. Although these studies are in progress, we have confirmed that dystrophin is expressed in a mosaic pattern,[54] and are currently studying mechanisms regulating its expression and distribution. These mosaic cells are models of the situation that might arise in muscle tissue of patients treated by cell grafting or gene therapy. We hope that the biological information gained from such studies will be valuable in designing such therapeutic approaches. We have also accumulated data, similar to that mentioned above, that in the dog small diameter fibres are resistant to degeneration. In collaboration with investigators at the University of Pennsylvania we are also studying the bioenergetics of dystrophic muscle using phosphorus magnetic resonance spectroscopy (^{31}P-MRS). In *xmd* dogs Pi/PCr is elevated, as it is in man (W J Bank, personal communication). This abnormality is exacerbated by 'exercise' simulated by repetitive neural stimulation.

Results from studies such as these will undoubtedly increase our understanding of the pathogenesis of muscle injury in Duchenne-type muscular dystrophy and should suggest potential new approaches to therapy, which can in turn be tested in these models.

CONCLUSIONS AND SPECULATIONS

The situation regarding animal models of muscular dystrophy of the Duchenne type has changed as dramatically over the last few years as has that of Duchenne muscular dystrophy itself. We now have two models which are known to be homologous to DMD at the molecular level. Each of these models offers advantages and disadvantages. The disadvantage of the *mdx* mouse is that it develops little or no clinical disease, and lesions are relatively mild and non-progressive compared to DMD. However, mice are easily managed and homozygous *mdx* mice can be raised and bred in relatively large numbers. Although dogs are more difficult than mice to keep, the advantage of the *xmd* dog is that the disease closely parallels the human condition clinically and pathologically. In particular dystrophic dogs develop progressive fibrosis of muscle and clinical cardiomyopathy very similar to that seen in Duchenne patients. Clearly the dog model will be particularly useful for investigation of the pathogenesis of these lesions and for therapeutic trials. Both models will be useful for understanding the pathologic impacts of the absence of dystrophin from muscle cells, and it will be useful to validate observations from one model in the other. It will also be important to understand why lack of dystrophin leads to devastating disease in man and dog, but relatively mild disease in the mouse.

Equally important is the question of why necrosis is self-limiting in the mouse. Does the persistence of central nuclei indicate cytoskeletal differences that compensate for the lack of dystrophin? Finally, what does dystrophin do? Investigations of dystrophic muscle lacking dystrophin using structural, biochemical and physiological techniques should shed some light on these questions. We can look forward to considerable progress in understanding the pathobiology of muscular dystrophy in the next few years.

ACKNOWLEDGEMENTS

The author wishes to thank his colleagues Eleene Gallagher, Beth Valentine, Nena Winand and Tony Perdue for their tireless efforts in studying the *xmd* dog and for comments on this manuscript. The work on the *xmd* dog described herein is supported by grants from the Muscular Dystrophy Association and the Cornell Biotechnology Program.

REFERENCES

1 Mrak RE. Muscular dystrophy. In: Mrak RE, ed. Muscle Membranes in Diseases of Muscle. Boca Raton: CRC Press, 1985: pp. 23–80

2 Mendell JR, Higgins R, Sahenk Z, Cosmos E. Relevance of genetic animal models of muscular dystrophy to human muscular dystrophies. Ann NY Acad Sci 1979; 317: 409–429

3 Ohno S. Ancient linkage groups and frozen accidents. Nature 1973; 244: 259–262

4 Bulfield G, Siller WG, Wight PAL, Moore KJ. X chromosome-linked muscular dystrophy (*mdx*) in the mouse. Proc Natl Acad Sci USA 1984; 81: 1189–1192

5 Chamberlain JS, Pearlman JA, Muzny DM, et al. Expression of the murine Duchenne muscular dystrophy gene in muscle and brain. Science 1988; 239: 1416–1418

6 Torres LFB, Duchen LW. The mutant *mdx*: inherited myopathy in the mouse. Morphological studies of nerves, muscles and end-plates. Brain 1987; 110: 269–300

7 Dangain J, Vrbova G. Muscle development in mdx mutant mice. Muscle Nerve 1984; 7: 700–704

8 Tanabe Y, Esaki K, Nomura T. Skeletal muscle pathology in X chromosome-linked muscular dystrophy (*mdx*) mouse. Acta Neuropathol (Berl) 1986; 69: 91–95

9 Carnwath JW, Shotton DM. Muscular dystrophy in the *mdx* mouse: histopathology of the soleus and extensor digitorum longus muscles. J Neurol Sci 1987; 80: 39–54

10 Coulton GR, Morgan JE, Partridge TA, Sloper JC. The mdx mouse skeletal muscle myopathy: I. A histological, morphometric and biochemical investigation. Neuropathol Appl Neurobiol 1988; 14: 53–70

11 Karpati G, Carpenter S, Prescott S. Small-caliber skeletal muscle fibers do not suffer necrosis in mdx mouse muscular dystrophy. Muscle Nerve 1988; 11: 795–803

12 Glesby MJ, Rosenmann E, Nylen EG, Wrogemann K. Serum CK, calcium, magnesium, and oxidative phosphorylation in mdx mouse muscular dystrophy. Muscle Nerve 1988; 11: 852–856

13 Anderson JE, Ovalle WK, Bressler BH. Electron microscopic and autoradiographic characterization of hindlimb muscle regeneration in the mdx mouse. Anat Rec 1987; 219: 243–257

14 Woo M, Tanage Y, Ishii H, Nonaka I, Yokoyama M, Esaki K. Muscle fiber growth and necrosis in dystrophic muscles: a comparative study between *dy* and *mdx* mice. J Neurol Sci 1987; 82: 111–122

15 Coulton GR, Curtin NA, Morgan JE, Partridge TA. The mdx mouse skeletal muscle myopathy: II. Contractile properties. Neuropathol Appl Neurobiol 1988; 14: 299–314

16 Bridges LR. The association of cardiac muscle necrosis and inflammation with the degenerative and persistent myopathy of MDX mice. J Neurol Sci 1986; 72: 147–157

17 Kunkel LM, Monaco AP, Middlesworth W, Ochs HD, Latt SA. Specific cloning of DNA fragments absent from the DNA of a male patient with an X chromosome deletion. Proc Natl Acad Sci USA 1985; 82: 4778–4782

18 Monaco AP, Neve RL, Colletti-Feener C, Bertelson CJ, Kurnit DM, Kunkel LM. Isolation of candidate cDNAs for portions of the Duchenne muscular dystrophy gene. Nature 1986; 323: 646–650

19 Hoffman EP, Brown RH Jr, Kunkel LM. Dystrophin: the protein product of the Duchenne muscular dystrophy locus. Cell 1987; 51: 919–928

20 Hoffman EP, Monaco AP, Feener CC, Kunkel LM. Conservation of the Duchenne muscular dystrophy gene in mice and humans. Science 1987; 238: 347–350

21 Avner P, Amar L, Arnaud D, Hanauer A, Cambrou J. Detailed ordering of markers localizing to the Xq26-Xqter region of the human X chromosome by the use of an interspecific *Mus spretus* mouse cross. Proc Natl Acad Sci USA 1987; 84: 1629–1633

22 Chamberlain JS, Grant SG, Reeves AA, et al. Regional localization of the murine Duchenne muscular dystrophy gene on the mouse X chromosome. Somatic Cell Mol Genet 1987; 13: 671–678

23 Brockdorff N, Cross GS, Cavanna JS, et al. The mapping of a cDNA from the human X-linked Duchenne muscular dystrophy gene to the mouse X chromosome. Nature 1987; 328: 166–168

24 Heilig R, Lemaire C, Mandel J-L, Dandolo L, Amar L, Avner P. Localization of the region homologous to the Duchenne muscular dystrophy locus on the mouse X chromosome. Nature 1987; 328: 168–170

25 Ryder-Cook AS, Sicinski P, Thomas K, et al. Localization of the *mdx* mutation within the mouse dystrophin gene. EMBO J 1988; 7: 3017–3021

26 Sugita H, Arahata K, Ishiguro T, et al. Negative immunostaining of Duchenne muscular dystrophy (DMD) and *mdx* muscle surface membrane with antibody against synthetic peptide fragment predicted from DMD cDNA. Proc Jpn Acad [B] 1988; 64: 37–39

27 Arahata K, Ishiura S, Ishiguro T, et al. Immunostaining of skeletal and cardiac muscle surface membrane with antibody against Duchenne muscular dystrophy peptide. Nature 1988; 333: 861–863

28 Valentine BA, Cooper BJ, Cummings JF, de Lahunta A. Progressive muscular dystrophy in a golden retriever dog: light microscopic and ultrastructural features at 4 and 8 months. Acta Neuropathol (Berl) 1986; 71: 301–310

29 Meier H. Myopathies in the dog. Cornell Vet 1958; 48: 313–330

30 de Lahunta A. Veterinary Neuroanatomy and Clinical Neurology. Philadelphia: Saunders, 1977; pp. 84–85

31 Kornegay JN, Tuler SM, Miller DM, Levesque DC. Muscular dystrophy in a litter of golden retriever dogs. Muscle Nerve 1988; 11: 1056–1064

32 Cardinet GH, Holliday TA. Neuromuscular diseases of domestic animals: a summary of muscle biopsies from 159 cases. Ann NY Acad Sci 1979; 317: 290–311

33 Cooper BJ, Valentine BA, Wilson S, Patterson DF, Concannon PW. Canine muscular dystrophy: Confirmation of X-linked inheritance. J Hered 1988; 79: 405–408

34 Valentine BA, Cooper BJ, de Lahunta A, O'Quinn R, Blue JT. Canine X-linked muscular dystrophy. An animal model of Duchenne muscular dystrophy: Clinical studies. J Neurol Sci 1988; 88: 69–81

35 Cooper BJ, Winand NJ, Stedman H, et al. The homologue of the Duchenne locus is defective in X-linked muscular dystrophy of dogs. Nature 1988; 334: 154–156

36 Bodensteiner JB, Engel AG. Intracellular calcium accumulation in Duchenne dystrophy and other myopathies: a study of 567,000 muscle fibers in 114 biopsies. Neurology 1978; 28: 439–446

37 Mokri B, Engel AG. Duchenne dystrophy: electron microscopic findings pointing to a basic or early abnormality in the plasma membrane of the muscle fiber. Neurology 1975; 25: 1111–1120

38 Valentine BA, Cooper BJ, Gallagher EA. Intracellular calcium in canine muscle biopsies. J Comp Pathol 1988 (In Press)

39 Frankel KA, Rosser RJ. The pathology of the heart in progressive muscular dystrophy: epimyocardial fibrosis. Hum Pathol 1976; 7: 375–386

40 Perloff JK, de Leon AC, O'Doherty D. The cardiomyopathy of progressive muscular dystrophy. Circulation 1966; 33: 625–648

41 Koenig M, Hoffman EP, Bertelson CJ, Monaco AP, Feener C, Kunkel LM. Complete cloning of the Duchenne muscular dystrophy (DMD) cDNA and preliminary genomic organization of the DMD gene in normal and affected individuals. Cell 1987; 50: 509–517

42 Shibuya S, Wakayama Y. Freeze-fracture studies of myofiber plasma membrane in X chromosome-linked muscular dystrophy (mdx) mouse. Acta Neuropathol (Berl) 1988; 76: 179–184

43 Wakayama Y, Okayasu H, Shibuya S, Kumagai T. Duchenne dystrophy: reduced density of orthogonal array subunit particles in muscle plasma membrane. Neurology 1984; 34: 1313–1317

44 Fischbeck KH, Bonilla E, Schotland DL. Distribution of freeze-fracture particle sizes in Duchenne muscle plasma membrane. Neurology 1984; 34: 534–535

45 Bonilla E, Fischbeck K, Schotland DL. Freeze-fracture studies of muscle caveolae in human muscular dystrophy. Am J Pathol 1981; 104: 167–173

46 Schotland DL, Bonilla E, Wakayama Y. Freeze fracture studies of muscle plasma membrane in human muscular dystrophy. Acta Neuropathol (Berl) 1981; 54: 189–197

47 Sawada H, Tsuji S, Kusumoto S, Doi Y, Matsushita H. Preclinical increase in activity of muscle microsomal trypsin-like protease in murine muscular dystrophy, C57BL/10-mdx. FEBS Lett 1986; 199: 193–197

48 Sano M, Wada Y, Ii K, Kominami E, Katunuma N, Tsukagoshi H. Immunolocalization of cathepsins B, H and L in skeletal muscle of X-linked muscular dystrophy (mdx) mouse. Acta Neuropathol (Berl) 1988; 75: 217–225

49 Christie KN. Chymostatin has no apparent beneficial effect on muscular dystrophy in the mdx mouse. J Neurol Sci 1988; 84: 341

50 Turner PR, Westwood T, Regan CM, Steinhardt RA. Increased protein degradation results from elevated free calcium levels found in muscle from mdx mice. Nature 1988; 335: 735–738

51 Schmalbruch H. Segmental fiber breakdown and defects of the plasmalemma in diseased human muscles. Acta Neuropathol (Berl) 1975; 33: 129–141

52 Rowland LP. Biochemistry of muscle membranes in Duchenne muscular dystrophy. Muscle Nerve 1980; 3: 3–20

53 Jones GE, Witkowski JA. Membrane abnormalities in Duchenne muscular dystrophy. J Neurol Sci 1983; 58: 159–174

54 Cooper BJ, Winand NJ, Sylvester JE, et al. Molecular relationship of canine X-linked muscular dystrophy to Duchenne muscular dystrophy of man. 4th Int Congress Cell Biol 1988; 424. Abstract P13.3.10

British Medical Bulletin (1989) Vol. 45, No. 3, pp. 719–744

Carrier detection and prenatal diagnosis in Duchenne and Becker muscular dystrophy

S V Hodgson
M Bobrow
Paediatric Research Unit, Division of Medical and Molecular Genetics, United Medical and Dental Schools of Guy's and St. Thomas's Hospitals, Guy's Hospital, London, UK

Estimating carrier risks for female relatives of Duchenne (DMD) and Becker (BMD) dystrophy sufferers depends upon calculation of segregational risks, supplemented by enzyme tests which show considerable overlap between carrier and control data. Linkage analysis has substantially increased the accuracy of segregational risk estimation, but a small error rate is still inherent when interpreting results, owing to recombination between the mutation causing the disease, and the marker used. It also requires family studies, which may be difficult to complete. The presence of intragenic DNA deletions in about half of D/BMD boys, allows direct detection of the D/BMD mutation, and is a powerful diagnostic tool. These techniques can be used for both prenatal diagnosis and carrier detection.

There is still no treatment for Duchenne and Becker muscular dystrophy. Genetic counselling is therefore very important to females at risk of carrying these conditions. They should be counselled before they undertake pregnancies, so that they can make informed decisions about childbearing, and whether they want prenatal diagnosis. Unfortunately, as yet there is no definitive carrier test. Recent progress in elucidating the molecular basis of these disorders[1-5] has provided new techniques for carrier detection and prenatal diagnosis.

0007–1420/89/0045–0719/$10.00

The first requirement for genetic counselling is a correct diagnosis in the affected male(s). This is based on clinical evaluation, serum creatine phosphokinase (CPK), electromyography (EMG), muscle biopsy, and muscle ultrasound examination. The diagnosis in about 50% of cases can be confirmed by showing a molecular deletion within the D/BMD gene. Quantitative or qualitative abnormalities of dystrophin[4] in muscle biopsy may soon become the definitive diagnostic text.

Once diagnosis is established, the carrier risk in a specified female relative may be assessed on the basis of pedigree analysis, biochemical tests of carrier status, and linkage analysis. We will review the interactive roles of these various methods of carrier risk assessment, give some examples of common clinical problems, and discuss the application of new genetic techniques to prenatal diagnosis.

PEDIGREE ANALYSIS

DMD and BMD are X-linked recessive conditions. The disease is fully expressed in males, and transmitted by females. Each son of an obligate carrier (usually a woman with an affected son and an antecedent relative) has a 50% risk of the disease, and each daughter has a 50% risk of being a carrier. The carrier risk of a woman related to an affected boy can be calculated using Bayes's theorem, which allows incorporation of information about her normal male relatives.[6]

Sporadic cases: consideration of mutation rates

Assessing the carrier risk in the mother of a sporadically affected boy is the most common counselling question in this disease. Some mothers are carriers who by chance have no affected relatives, and others are non-carriers who have had an affected child owing to a new mutation.

When estimating carrier risk for the mother of an isolated case of DMD, it is important to consider the sex-specific rate of new mutations in this condition. Haldane[7] calculated that, since DMD is a genetic lethal trait, two thirds of the mothers of all affected boys should be carriers, provided the mutation rate is equal in male and female gametes. Whether these mutation rates are in fact equal is of more than academic importance. If most mutations occurred in male gametes, the risk estimate for the mothers of isolated cases would be higher.[8]

Certain studies apparently detect a higher proportion of carrier women in mothers of isolated cases than would be expected. Roses et al.[9] found a significant increase in peak II endogeneous phosphorylation of the red cell membrane in DMD patients, and in all of 14 possible carrier mothers. Pickard et al.[10] found reduced lymphocyte capping in boys with DMD, and also in 84% of mothers of isolated cases.[11,12] However, other workers have not confirmed these results.[13-17] Bucher et al.,[18] assaying muscle ribosomal protein synthesis,[19,20] estimated that 77% of mothers of isolated cases were carriers.

Most authors using CPK and other methods of carrier detection agree that 2/3 of mothers of DMD boys are carriers.[21] This is generally supported by segregation analysis,[22-25] although Danieli and Barbujani[26] found a smaller than expected number of sporadic cases. Roses and his co-workers have presented data suggesting that a high proportion of mothers of DMD boys with deletions are also deletion carriers,[27] but preliminary linkage analysis[28,29] has been consistent with equal mutation rates in both sexes, although it is difficult to allow for ascertainment bias in analysing these data.[30,31] There is evidence that male mutation rates exceed female at the loci determining haemophilia A and Lesch-Nyhan syndrome.[32-34] The balance of opinion is currently that male and female germ cell mutation rates are equal for DMD,[35-37] but the question is not yet settled.

TESTS OF CARRIER STATUS

Historical review

Carrier diagnosis has rested on clinical examination, pedigree analysis, and various laboratory tests. About 10% of DMD carriers are manifesting—that is, they show some features of the disease.[38-41] In the 1950s, raised levels of serum CPK were noted in many obligate carriers of DMD.[42,43] This is the most widely accepted test, and is reviewed below.

Many abnormalities found in boys with DMD have been found to an intermediate degree in carriers, including abnormalities of the electrocardiogram[44-46] and EMG[47-49] and elevated serum aldolase,[50] carbonic anhydrase III,[51] pyruvate kinase,[52-55] myoglobin,[56-59] and haemopexin.[56] Other abnormalities reported are reduced muscle LDH-5 isoenzyme,[60,61] and increased intracellular calcium:phosphate ratio.[62] Erythrocyte surface deformation has been detected by scanning electron microscopy in carriers,[63,64] although this was not confirmed by other workers.[11] A decreased number of intramembrane particles in erthyrocytes of carriers has been reported using freeze-fracture analysis.[65]

About one third of female carriers show myopathic changes in muscle histology.[66-69] Electron microscopic[70-72] and histochemical abnormalities of muscle fibres have also been demonstrated.[73,74] Abnormal ultrasound and computer-aided tomographic (CT) findings are reported,[75-77] but are not sufficiently consistent to be used for carrier detection.[78]

Use of CPK estimations for carrier detection

The serum CPK has been most widely used for carrier detection in DMD and is thought to detect between 70–75% of carriers.[79-82]

Variability in different samples from the same subject has led to using the mean of three separate CPK estimations from each woman at risk.[83] The assay must be meticulously standardized for this specific purpose.[84,85] CPK in control females and obligate carriers forms two overlapping log-normal distributions.[86-88] A standard risk curve is constructed by each individual laboratory, showing the ratio of obligate carrier curve to normal control curve for each CPK value. From this can be read the probability that a woman with a particular mean CPK is a carrier (given prior carrier odds of 1:1[81,84,88]). These CPK-based estimates of carrier risk can be combined with pedigree data using Bayes's theorem.[87,89,90]

CPK levels in carriers tend to be higher before puberty, so testing is generally postponed until adulthood.[91,92] Some workers have suggested early teenage carrier testing, to increase test sensitivity.[92,93] CPK values in postmenopausal women have variously been found to be slightly raised,[91,94] lowered,[80,92,93,95-97] or unchanged.[98]

Attempts to increase test sensitivity by CPK estimation after physical exercise[88,94,99,100] have proved unreliable.[101,102] Pregnancy lowers serum CPK values in DMD carriers.[98,103-106] The menstrual cycle and the contraceptive pill have no effect on serum CPK values.[85,94,99]

CPK levels can be used for carrier detection in BMD in a similar manner, provided a standard curve for obligate carriers has been established.[107-111]

LINKAGE ANALYSIS

Estimates of serum CPK and other parameters do not on their own provide reliable carrier detection tests, because of their variability and the overlap between normal and carrier values. The discovery of polymorphic loci within or closely linked to D/BMD offers more accurate carrier risk prediction within families. Linkage analysis using information from such loci is now widely used for genetic counselling in DMD and BMD families.[112-115]

If a woman carrying the DMD mutation has different polymorphic marker alleles on her two X chromosomes (i.e. she is heterozygous), and the allele inherited by her affected son is ascertained, another of her children inheriting the same allele will also have inherited the DMD mutation, unless recombination has occurred so that the marker and disease loci are no longer carried on the same X chromosome. The recombination fraction Θ describes how often such recombination occurs. The closer the marker and disease loci are to each other, the smaller Θ is. The unusual size and high mutation rate of D/BMD pose specific problems. Many different individual mutations occur, the site of a particular mutation is frequently unknown, and considerable recombination can occur within the large gene. The recombination frequency between the pERT and XJ loci and DMD/BMD is about 5%,[116,117] even though they all lie within the same gene.

For genetic counselling, the usefulness of a linked locus depends upon the recombination fraction, and also on the frequency of heterozygosity at the marker locus. If the rare allele of a di-allelic polymorphism occurs at low frequency, most females tested will be homozygous for the common allele, which will be 'uninformative' for counselling. This can be overcome by using several different polymorphisms.

The probes most widely used for counselling are the pERT series[118] and the pXJ sequences.[119] Probes in regular use in our clinical laboratory are listed in Table 1.

The use of 'bracketing' markers on both sides of the DMD locus provides more accurate information than a single marker locus alone, because a double recombination would need to occur within the region defined by the bracketing loci to cause erroneous prediction.[112,120] Where recombination has occurred between bracketing loci, interpretation is difficult because the exact position of the recombination relative to the D/BMD mutation in that particular family is not known. It is imprudent to give low risks in

Table 1 Probes in regular use in our clinical laboratory

		Θ used for counselling
Proximal	L1–28	0.20
	754	0.16
Intragenic	XJ1 (XJ1.1)	
	XJ1.2	
	XJ5	
	XJ8	
	HIP25	
	pERT 87–1	
	pERT 87–8	0.05
	pERT 87–15	
	pERT 87–30	
	J–Bir	
	J47	
	JMD–2	
	J66–HI	
	p20	
Distal	C7	0.12
	99.6	0.15
	RC 8	0.20

such a situation. Intragenic loci can be used as flanking markers if the position of the DMD mutation has been demonstrated by deletion mapping.

THE APPLICATION OF LINKAGE ANALYSIS

The prior probability that a woman is a carrier, modified by the number of her normal male relatives and combined with information from linkage and biochemical tests, is used in a Bayesian calculation to provide an integrated risk figure. The methodology is described by Emery,[37] and an example of such a calculation is shown in Figure 1.

Each time a chromosome passes from one generation to another, there is a possibility that recombination will take place between the marker and disease loci. Where linkage information is only available from one brother of the consultand, the possibility of recombination causing diagnostic error is 2Θ, as recombination could have occurred between the mother and the consultand, or between mother and affected son. This error rate is reduced if linkage information from two or more brothers is available, because this establishes 'phase' (which marker is travelling with the disease allele) on two independent observations.

These calculations can be complicated, particularly when deal-

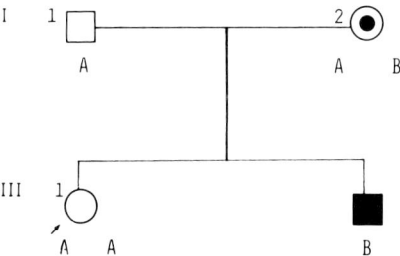

III.1's CPK gives 1:3 ratio of carrier:not carrier.

	B with dis:A with dis		B with dis:A with dis	
Linkage phase	1-θ	θ	1-θ	θ
consultand:		Carrier		Not Carrier
	θ	1-θ	1-θ	θ
CPK data	1	1	3	3
	2θ(1-θ)		3(1-θ)²	3θ²

Carrier risk:
$$\frac{2\theta(1-\theta)}{2\theta(1-\theta) + 3\theta^2 + 3(1-\theta)^2} = 3.38\% \text{ for } \theta = 0.05$$

Fig. 1 Risk calculation using linkage analysis.

ing with data from multiple linked markers, and may require computer analysis.[121,122]

Female relatives of affected boys present somewhat different problems, according to the relationship.

Mothers

If the mother of an affected boy has another affected relative, she is an obligate heterozygote. Linkage analysis cannot alter her carrier risk, although it may permit prenatal diagnosis (see section on *Prenatal diagnosis*).

If the mother of an isolated case of DMD also has a normal son(s), the presence of the same X chromosome in both normal and affected boys will increase the probability that this is a new mutation, and that she is therefore not a carrier (*see* Fig. 2). If they inherited different X chromosomes, her carrier risk would be increased.

The composite risk to the mother of a sporadic case of DMD will be considerably influenced by her CPK results, and by the number of her normal sons or brothers.

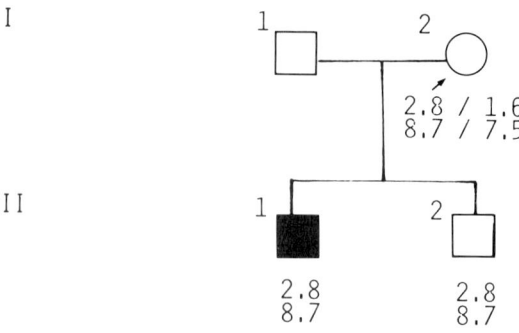

PROBES: 2.8 / 1.6 = PERT 87-1
8.7 / 7.5 = PERT 87-15

Fig. 2 Reduced carrier risk estimate in the mother of an isolated case of DMD, where her normal and affected sons have inherited the same alleles.

Sisters

The carrier risk of an affected boy's sister is increased if their mother has passed on the same allele to both of them at a locus linked to D/BMD (*see* Fig. 3). Knowing which allele an unaffected brother has inherited from his heterozygous mother can provide 'phase' information for his sister, even when her affected brother has died.

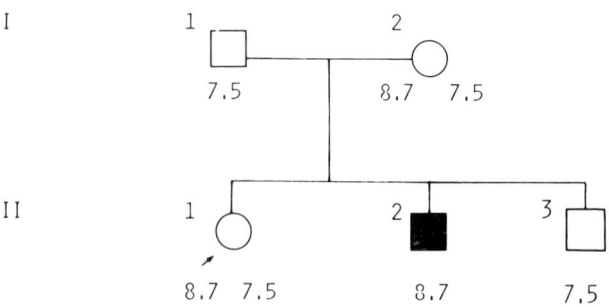

PROBE : 8.7 / 7.5 = PERT 87-1

Fig. 3 The mother has passed the same allele to the affected boy and his sister.

More distant female relatives

If extended family studies show that the DMD sufferer has inherited alleles at loci linked to D/BMD from his grandfather, the D/BMD mutation probably occurred in the germ cell line of the boy's mother, or of her father. This greatly reduces the grandmother's carrier risk, and the risks for maternal aunts are correspondingly reduced (*see* Fig. 4). Where the grandmother might be a carrier, linkage analysis can be used for the aunts as it is for sisters of the index case.

Information about the genotype of a woman's father may be essential for counselling—for instance, when she and her mother are both heterozygous at the informative locus (*see* Fig. 3). Wrong paternity information can lead to inappropriate deductions as to which maternal allele the consultand has inherited. Where indicated, paternity can be checked using hypervariable 'minisatellite' probes.[123]

THE USE OF DELETIONS IN COUNSELLING

The inaccuracy of linkage analysis, owing to the possibility of recombination between a marker and the D/BMD mutation, may be circumvented if an intragenic deletion is detected in the affected boy. These deletions are (with some exceptions)[124,125] the disease-causing mutations and are therefore precise predictors of disease.

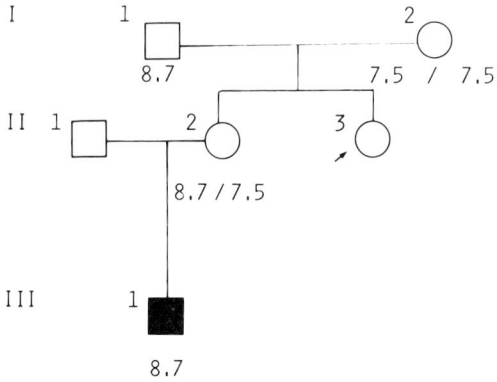

PROBE: PERT 87-1

Fig. 4 The affected boy has inherited alleles linked to *DMD* from his grandfather; this reduces the carrier risks of the maternal aunt and grandmother.

The pERT and XJ1.1 clones detect intragenic deletions in about 10% of D/BMD patients.[126-128] Use of multiple genome probes increases the proportion of cases showing deletions to over 25%.[126,129,130] cDNA clones representing the coding regions (exons) of D/BMD detect deletions in about 50% of affected boys.[3,131] Pulse field analysis also detects these deletions.[132] If a sporadic case has a DNA deletion, and his mother is found to be heterozygous at a locus within the region deleted in her son, her carrier risk estimate is reduced, as she is not carrying the deletion (*see* Fig. 5; but see also *Germinal mosaicism* below). The same would apply, with even greater certainty, to sisters and other female relatives.

Partial gene duplications have also been reported in DMD/BMD.[132-134]

Limitations and problems in the interpretation of deletion data

Detection of deletion carriers

It is technically difficult to determine whether a woman is homozygous at a locus, or is a deletion heterozygote and therefore hemizygous for the site(s) concerned. Quantitative distinction between one and two copies of a locus on Southern blots is generally not considered robust enough for regular clinical use. Demonstration of an abnormal-sized 'junction' band is extremely valuable; such markers are seen in a few families, but are more frequently shown on field inversion gel electrophoresis.[132]

Where a daughter fails to inherit a maternal allele (e.g. Fig. 6), the transmission of a deletion from mother to daughter can be inferred, implying that both are carriers; occasionally, this enables a diagnosis to be established in the absence of any informative males![126]

Germinal mosaicism

Women heterozygous for loci deleted in their sons have, on several occasions, passed on a deletion to more than one of their offspring.[135-137] This phenomenon has also been reported in the offspring of an unaffected male.[138] The suggested cause is germline cell mosaicism, although this is not yet formally demonstrated. Bakker et al.[135] estimate the frequency of this phenome-

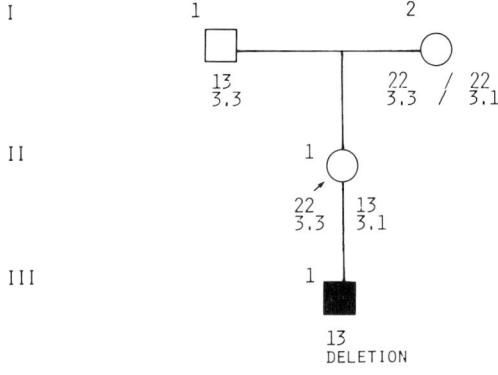

Fig. 5 The mother is heterozygous at the site for which her affected son is deleted, and does not therefore carry the deletion, so that her carrier risk is reduced.

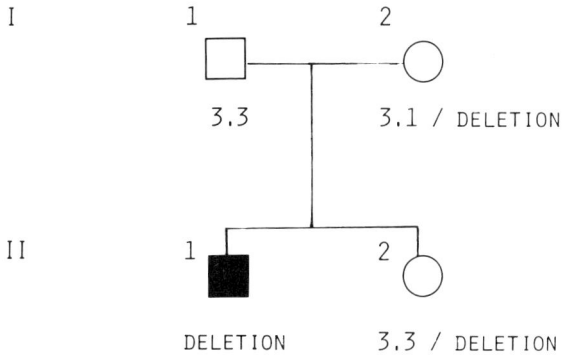

Fig. 6 The sister of the affected boy appears not to have inherited a maternal allele; she is deduced to have inherited the deleted allele and therefore has a high carrier risk.

non as around 10%, but this is based on very small numbers and awaits confirmation. Until this finding is clarified, it is necessary to exercise caution in reassuring mothers of boys with deletions, particularly since a deletion allows very accurate prenatal diagnosis. There is no reason to be similarly conservative when counselling other female relatives. It is as yet not possible to determine whether similar situations exist in cases with no detectable deletion.

Non-pathogenic deletions

There have been instances in which a similar deletion has been found in both affected males and unaffected males in the same family.[124,125] This very unusual finding is a potential cause of diagnostic error, which can be avoided by making certain that deletions thought to cause the disease are not entirely within an intron.[125]

GENERAL CONCLUSIONS ABOUT CARRIER RISK COUNSELLING

Uptake of genetic counselling services is high, and the more accurate risk estimates attainable using DNA analysis are welcomed by affected families.[114] No counselling can be initiated without careful prior discussions, explaining the accuracy and limitations of the tests. It is also important to offer counselling to the extended family. An average of 4–6.5 blood samples per family is needed to counsel the close family of an index case of DMD.

Blood samples should be obtained as early as possible from index cases and other critical family members, and 'banked' for possible later use.

Conclusions using linkage analysis in general correlate well with carrier risk estimates based on CPK and pedigree data.[114,120] The combined results tend to shift risk estimates towards the extremes of the range of risk, leaving fewer women in the ambiguous middle range between 20% and 80% (*see* Fig. 7 [A–C]). DNA analysis is less helpful for mothers of isolated cases, as it seldom helps to distinguish carrier mothers from those whose sons are the result of a new mutation.

Women find that clarification of their level of risk makes decisions about childbearing much easier. Many women are now having families who were not prepared previously to run the risks of having affected children.

PRENATAL DIAGNOSIS

Introduction

Before the advent of linkage techniques, carrier women could choose fetal sexing and termination of all male pregnancies, rather than risk an affected child; many found this so distressing that they refrained from having children. Improved carrier detection allows more informed decisions to be made about childbearing. There is a wide variation in the level of carrier risk at which childbearing without prenatal diagnosis is acceptable, but a pregnancy risk of 5% or less is acceptably low for many women.[120,139]

Historical review

Histological abnormalities have been demonstrated in muscle from 16 week fetuses with probable DMD.[140-142] Raised levels of CPK in fetal blood were considered to indicate an affected fetus[143,144] but this was found to be unreliable,[145,146] and is no longer used. Other tests (e.g. fetal plasma carbonic anhydrase III and myoglobin levels) for fetal diagnosis were attempted with variable and unsatisfactory results.[51,147-149]

Obstetric techniques

Until recently, amniocentesis at 16 weeks gestation was the method commonly used for prenatal sampling. Culture of the amniocytes is required to obtain sufficient material for karyotyping and DNA studies, and the result is not available for several weeks. The procedure carries a 0.5–1% risk of a miscarriage, but if termination of pregnancy is required, it must be performed in the late second trimester, and many women experience depression afterwards.[150]

The development of chorionic villus sampling has allowed diagnostic procedures to be performed from about 10 weeks of gestation. Sufficient fetal material can be obtained by this procedure for rapid sexing, and preparation of DNA for linkage studies, making results available within the first trimester of pregnancy.[139,151] Estimates of fetal loss due to chorion biopsy vary widely;[152,153] the results of randomized studies are still awaited, but figures between 1 and 3% are usually quoted. Most women in high-risk groups consider the advantages of chorionic villus biopsy outweigh the higher risk.

Current techniques of prenatal testing

Before undertaking prenatal diagnosis, the woman's carrier risk must be fully evaluated. A reduction in risk may alter her desire for testing. If tests are required, linkage analysis can usually

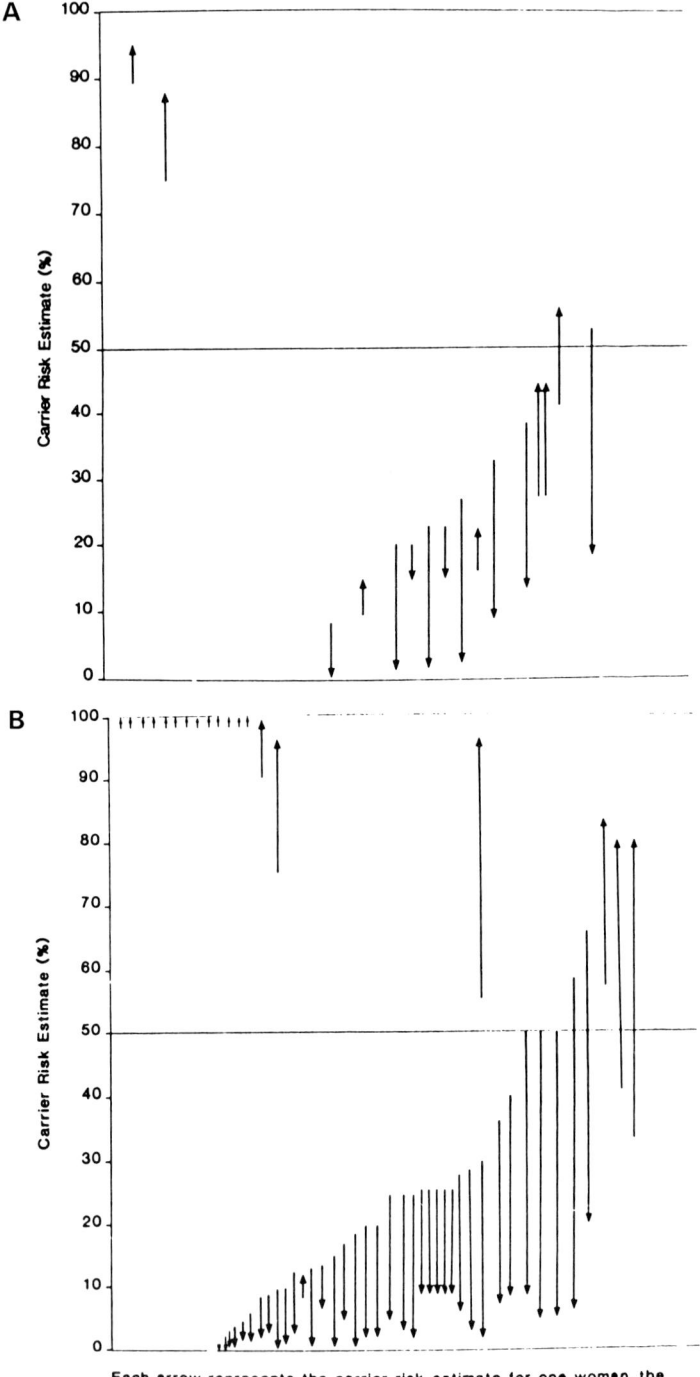

Each arrow represents the carrier risk estimate for one woman, the arrow starts at her risk using pedigree and CPK data only and ends at her risk incorporating linkage analysis

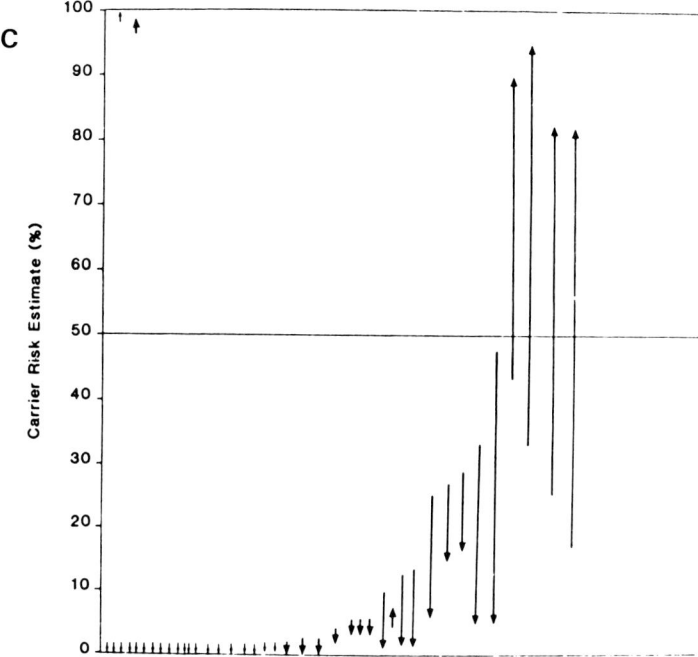

Fig. 7 Alteration of carrier risk estimates, using linkage analysis [A] for mothers of sporadic cases of DMD; [B] for sisters of index cases; and [C] for other female relatives of index cases.

distinguish between a male fetus at very low risk of being affected, and one at a higher risk. The high risk maternal chromosome confers on the fetus a risk of being affected equal to the maternal carrier risk, less the probability of recombination.

In a familial case, the mother's paternal X chromosome must be low-risk; all her carrier risk will be on her maternal X. If a male fetus has inherited the opposite allele from the affected brother (*see* Fig. 8), it will be at low risk of being affected. If a woman is the mother of an isolated case, phase information to identify her high-risk X chromosome can only be obtained from generations below her.

In many cases, prenatal exclusion of a high risk to the fetus requires blood samples from very few people—e.g. from the mother and one son.

Where an affected boy has been found to bear a deletion, this forms a precise marker for future pregnancies in that family (see Fig. 9).[131,139,154]

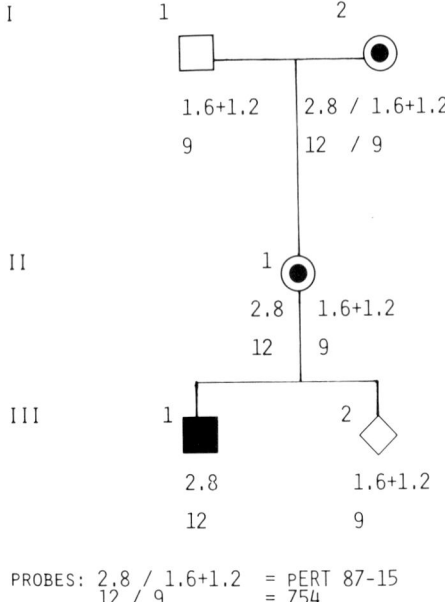

Fig. 8 The fetus carries the low-risk X chromosome inherited from its grandfather, opposite to that inherited by its affected brother.

Problems with diagnosis

The problems in using linkage analysis for prenatal diagnosis are the same as for genetic counselling—the mother must be informative at a suitable locus linked to D/BMD; family studies are needed, preferably before prenatal testing,[139] and the possibility of recombination between the informative marker and D/BMD remains a potential cause of erroneous diagnoses. It is essential that the pregnant woman and her husband receive accurate genetic counselling, and understand both the risks of diagnostic error and of the sampling procedure.

Confirmation of diagnosis

It is unfortunately impossible to confirm the diagnosis of dystrophy in a fetus on histological examination of products of first-trimester termination. DNA results can be confirmed. Dystrophin assays may in future allow direct fetal confirmation. This is important for audit purposes, but may also very substantially alter carrier risk estimates, particularly in relatives of isolated cases.

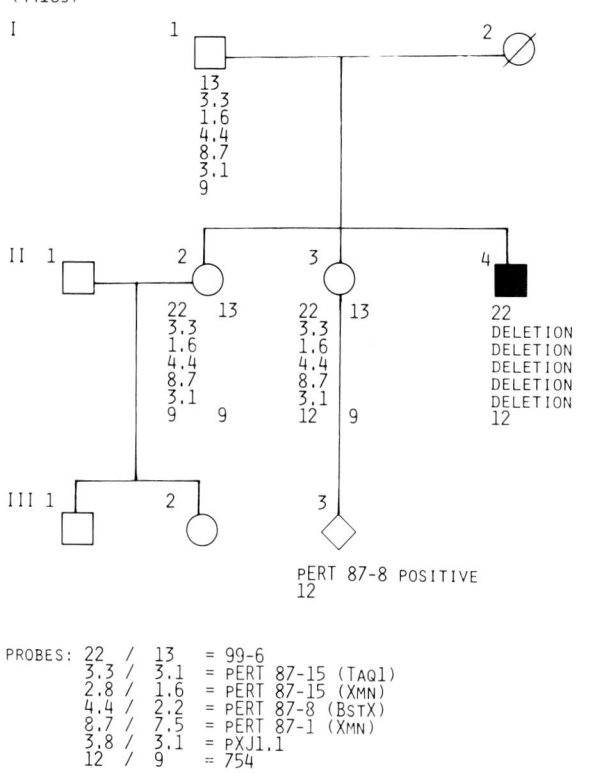

Fig. 9 The fetus does not carry the deletion found in its affected uncle, and is therefore at a very low risk of being affected.

Where a low-risk pregnancy continues to term, CPK testing of the baby can be done at about 6 weeks of age. Testing at birth is not advisable, as many neonates have raised CPK levels.[155-159]

CONCLUSION

Effects of genetic counselling

A genetic counselling programme enables the birth of second affected cases of DMD to be prevented.[160,161] In the past this was achieved when women with a high carrier risk avoided pregnancy or terminated all male pregnancies. The new techniques allow more accurate carrier detection. They also enable first trimester prenatal diagnosis to be offered to most women at risk, thereby allowing them the possibility of planning a family with reasonable confidence. If all women tested were definite carriers, 50% of male pregnancies would be found to be at high risk. In practice, many mothers of sporadic cases are not definite carriers, and therefore, more pregnancies will prove to be at low risk, and will continue, than will prove to be at high risk.

Because of the need to find an informative locus in the woman herself, to carry out family studies and, if possible, to screen an affected male relative for a deletion, it is very advantageous if she is referred well before embarking on pregnancy. Rapid analysis in a family where a woman presents already pregnant may lead to the use of suboptimal markers; it increases laboratory costs, and lessens the possibility of reducing carrier risk in the pregnant woman before prenatal diagnosis is undertaken. Most important, it often allows inadequate time for counselling, and for a calm and unpressured choice between available reproductive options.

Costs

The cost of DNA analysis is estimated to be £300–£350 per person, or £1 700–£2 300 per family. This must be a worthwhile expenditure, in view of the suffering caused by the disease, and the expense of caring for affected boys.[162]

THE FUTURE

Increasing knowledge of the D/BMD gene and its product will lead to continued improvements in carrier and prenatal detection. Direct demonstration of specific mutations, such as deletions, offers great precision, but is complicated by the wide variety of different mutations causing these conditions. The 50% of mutations which are not major deletions remain to be defined.

Since deletion screening is the most accurate method of prenatal diagnosis, their rapid detection using polymerase chain reaction (PCR)[163] will greatly facilitate prompt counselling in new families. Analysis of as few as 6 to 8 deletion-prone sites will allow very rapid detection of the great majority of deletions.

Because up to two thirds of cases of DMD are the first in their families, and no precise and direct screening test is yet possible for affected fetuses, or for carrier mothers without a family history, boys will continue to be born with this distressing disease. It is only in the prevention of secondary cases that these new techniques are currently helpful. However, as knowledge of the nature of the DMD gene and its product increases, we hope to understand the pathogenesis of the disorder further, and to develop more accurate methods for screening and prenatal diagnosis. The possibility of treatment always remains the ultimate goal.

ACKNOWLEDGEMENTS

Miss E. Manners gave invaluable help in collating the material for this paper, and in preparing the manuscript.

REFERENCES

1 Monaco AP, Neve RL, Colletti-Feener C, Bertelson CJ, Kurnit DM, Kunkel LM. Isolation of candidate cDNA for portions of the DMD gene. Nature 1986; 323: 646–650

2 Burghes AHM, Logan C, Hu X, Belfall B, Worton RG, Ray PN. A cDNA clone from the D/BMD gene. Nature 1987; 328: 434–437

3 Koenig M, Hoffman EP, Bertelson CJ, Monaco AP, Feener C, Kunkel LM. Complete cloning of the Duchenne muscular dystrophy (DMD) cDNA and preliminary genomic organization of the DMD gene in normal and affected individuals. Cell 1987; 50: 509–517

4 Hoffman EP, Fischbeck KH, Brown RH et al. Characterisation of dystrophin in muscle-biopsy specimens from patients with Duchenne's or Becker's muscular dystrophy. New Engl J Med 1988; 318: 1363–1368

5 Zubrzycka-Gaarn EE, Bulman DE, Karpati G et al. The DMD gene product is localized in sarcolemma of human skeletal muscle. Nature 1988; 333: 466–469

6 Emery AEH. Methodology in Medical Genetics—an introduction to statistical methods. Edinburgh: Churchill Livingstone, 1976: 92–95

7 Haldane JBS. The rate of spontaneous mutation of a human gene. J Genet 1935; 31: 317–326

8 Vogel F, Rothenberg R. Spontaneous mutations in Man. In: Harris H, Hirschhorn L, eds. Advances in Human Genetics 5. New York: Plenum Press, 1985: 223–318

9 Roses AD, Herbstreith M, Metcalf B, Appel SH. Increased phosphorylated components of erythrocyte membrane spectrin band II with reference to Duchenne dystrophy. Nature 1976; 254: 350–351

10 Pickard NA, Gonemer HD, Verril H et al. Systemic membrane defect in the proximal muscular dystrophies. N Engl J Med 1978; 299: 841–846

11 Verrill HL, Pickard NA, Gruemer HD. Diminished cap formation in lymphocytes from patients and carriers of Duchenne muscular dystrophy. Clin Chem 1977; 23: 2341–2343

12 Ho AD, Reitter B, Stojakowits S, Fiehn W, Weisser J. Capping of lymphocytes for carrier detection in Duchenne muscular dystrophy: technical problems and a review of the literature. Eur J Paediatr 1980; 134: 211–216

13 Burt D, Emery AEH. Serum LDH-5 in carriers of DMD. Neurology 1972; 29: 239–241

14 Gershwin ME, Taylor RG, Fowler WM Jr, Finlayson B. Failure to demonstrate abnormal lymphocyte capping in humans, mice and hamster with muscular dystrophy. Hum Genet 1979; 53: 113–114

15 Hauser S, Weiner H, Ault K, Unanue E. Lymphocyte capping in Duchenne muscular dystrophy. N Engl J Med 1979; 300: 361

16 Stern CMM, Kahan MC, Dubowitz V. Lymphocyte capping in DMD. Lancet 1979; I: 1300
17 Nicholson GA, Sugars J. An evaluation of lymphocyte capping in Duchenne muscular dystrophy. J Neurol Sci 1982; 53: 511–518
18 Bucher K, Ionasescu V, Hanson J. Frequency of new mutants amongst boys with DMD. Am J Med Genet 1980; 7: 27–46
19 Ionasescu V, Zellweger H, Conway TW. A new approach for carrier detection in DMD: protein synthesis of muscle polyribosomes in vitro. Neurology 1971; 21: 703–709
20 Ionasescu V, Burmeister L, Hanson J. Discriminant analysis of ribosomal synthesis findings in carrier detection of DMD. Am J Med Genet 1980; 5: 5–12
21 Percy ME, Chang LS, Murphy EG, Oss I, Verellen-Dumoulin C, Thompson MW. Serum CK and PX in DMD carrier detection. Muscle Nerve 1979; 2: 329–339
22 Thompson MW, Hutton EM. The occurrence of new mutants in linked recessive lethal disorders (abstract). In: Excerpta Medica International Congress series 397: Proceedings of the Vth International Congress on Human Genetics. Amsterdam: Excerpta Medica, 1976: pp. 195–196
23 Zatz M, Lange K, Spence MA. Frequency of DMD carriers. Lancet 1977; I: 759
24 Caskey CT, Nussbaum RL, Cohan LC, Pollack L. Sporadic occurrence of Duchenne muscular dystrophy: evidence for new mutation. Clin Genet 1980; 18: 329–341
25 Williams WR, Thompson MW, Morton NE. Complex segregation analysis and computer-assisted genetic risk assignment for Duchenne muscular dystrophy. Am J Med Genet 1983; 14: 315–333
26 Danieli GA, Barbujani G. Duchenne muscular dystrophy: frequency of sporadic cases. Hum Genet 1984; 67: 252–256
27 Roses AD. Mutants in Duchenne muscular dystrophy: implications for prevention. Arch Neurol 1988; 45: 84–85
28 Bobrow M, Walker A, Walton J. The parental origin of mutations causing Duchenne muscular dystrophy. Arch Neurol 1988; 45: 85–87
29 Müller CR, Grimm T. Estimation of the male to female ratio of mutation rates from the segregation of X-chromosomal DNA haplotypes in Duchenne muscular dystrophy families. Hum Genet 1986; 74: 181–183
30 Karel ER, te Meerman GJ, ten Kate LP. On the power to detect differences between male and female mutation rates for Duchenne muscular dystrophy, using classical segregation analysis and restriction fragment length polymorphisms. Am J Hum Genet 1986; 38: 827–840
31 te Meerman GJ, Karel ER, ten Kate LP. Ascertainment bias and power of procedures to estimate differences between male and female mutation rates. Hum Genet 1987; 75: 296
32 Francke U, Winter RM, Lin D, Bakay B, Seegmiller JE, Nyhan WL. Use of carrier detection tests to estimate male to female ratio of mutation rates in Lesch-Nyhan disease. In: Hook EB, Porter IH ed. Birth Defects Institute Symposia series: Population and Biological Aspects of Human Mutation. New York: Academic Press, 1981: pp. 17–30
33 Grimm T. The influence of half-chromatid mutations on the ratio of new mutations in lethal X-linked recessive disorders. Am J Hum Genet 1982; 34: 142–145
34 Winter RM, Tuddenham EDG, Goldman E, Matthews KB. A maximum likelihood estimate of the sex ratio of mutation rates in haemophilia A. Hum Genet 1983; 64: 156–159
35 Davie AM, Emery AEH. Estimation of proportion of new mutants among cases of Duchenne muscular dystrophy. J Med Genet 1978; 15: 339–345

36 Moser H. Duchenne muscular dystrophy: pathogenetic aspects and genetic prevention. Hum Genet 1984; 66: 17–40

37 Emery AEH. Oxford Monographs on Medical Genetics no. 15: Duchenne Muscular Dystrophy. Oxford: Oxford University Press, 1987

38 Moser H, Emery AEH. The manifesting carrier in Duchenne muscular dystrophy. Clin Genet 1974; 5: 271–284

39 Gomez MR, Engel AG, Dewald G, Peterson HA. Failure of inactivation of Duchenne dystrophy X chromosome in one of female identical twins. Neurology 1977; 27: 537–541

40 Yoshioka M. Clinically manifesting carriers in DMD. Clin Genet 1981; 20: 8–12

41 Olson BJ, Fenichel GM. Progressive muscle disease in a young woman with family history of DMD. Arch Neurol 1982; 39: 6, 378–389

42 Ebashi S, Toyokura Y, Homoi H, Sugita H. High creatine PK activity of sera of progressive muscular dystrophy patients. J Biochem (Tokyo) 1959; 46: 103

43 Dreifuss JC, Schapira G, Demos J. Etude de la creatine-kinase serique chez les myopathes et leur familles. Rev Franc Etudes Cliniques et Biologiques 1960; 5: 384

44 Mann O, De Leon AC, Perloff JK, Simanis J, Horrigan FD. Duchenne's muscular dystrophy: the ECG in female relatives. Am J Med Sci 1968; 255: 376–381

45 Emery AEH. Abnormalities of the ECG in female carriers of DMD. Br Med J 1969; 2: 418–420

46 Lane RJW, Gardner-Medwin D, Roses AD. EMG abnormalities in carriers of DMD. Neurology 1980; 30: 497–501

47 van den Bosch J. Investigations of the carrier state in the Duchenne type dystrophy. In: Research in Muscular Dystrophy. London: Pitman, 1963: 23–33

48 Moosa A, Brown BH, Dubowitz V. Quantitative electromyography—carrier detection in Duchenne type muscular dystrophy using a new automatic technique. J Neurol Neurosurg Psychiatr 1972; 35: 841–844

49 Hausmanowa-Petrusewicz I, Wierzbicka M, Joswik A, Szmidt-Salkowska E, Borkowska J. A nearest neighbour decision rule for EMG detection of carriers of DMD. Electromyogr Clin Neurophysiol 1982; 22: 445–457

50 Dreifuss JC, Schapira F. Biochemistry of hereditary myopathies II, 45C. Springfield, Ill: CC Thomas, 1962

51 Heath R, Carter ND, Jeffrey S, Edwards RJ, Watts DC, Watts RI. Evaluation of carrier detection of DMD using carbonic anhydrase III and CK. Am J Med Genet 1985; 21: 291–296

52 Alberts MC, Samaha FJ. Serum PK in muscular disease and carrier states. Neurology 1974; 24: 462–464

53 Zatz M, Shapiro LJ, Campion DS, Kaback MM, Otto PA. Serum pyruvate-kinase (PK) and creatine-phosphokinase (CPK) in female relatives and patients with X-linked muscular dystrophies (Duchenne and Becker). J Neurol Sci 1980; 46: 267–279

54 Zatz M, Passos MR, Rabbi-Bortolini E. Serum PK activity during pregnancy in potential carriers for DMD. Am J Med Genet 1983; 15: 149–151

55 Falcâo-Conceicâo DN, Gonçalves-Pimentel MM, Baptista ML, Ubatuba S. Detection of carriers of X-linked gene for DMD by levels of CK and PK. J Neurol Sci 1983; 62: 171–180

56 Adornato BT, Kagen LJ, Engel WK. Myoglobinaemia in Duchenne muscular dystrophy patients and carriers: a new adjunct to carrier detection. Lancet 1978; II: 499–501

57 Nicholson LVB. Serum myoglobin in Muscular dystrophy and carrier detection. J Neurol Sci 1981; 51: 411–426

58 Lössner J, Kuhn J, Ruchholtz V. Hämopexin und Muskeldystrophis. Zur

Karrierdiagnostik der Duchenne-Form. Psychiatr Neurol Med Psychol 1982; 34: 53–59

59 Percy ME, Pichora GA, Chang LS, Manchester KE, Andrews DF. Serum myoglobin in DMD carrier detection: a comparison with CK and hemopexin using logistic discrimination. Am J Med Genet 1984; 18: 279–287

60 Emery AEH. Electrophoretic pattern of LDH in carriers and patients with DMD. Nature 1964; 4923: 1044–1045

61 Somer H, Willner J, De Cresce RP, Willner JM. Duchenne carriers: LDH isoenzyme 5 in serum and muscle. Neurology 1980; 30: 206–209

62 Maunder-Sewry CA, Dubowitz V. Myonuclear calcium in carriers of Duchenne muscular dystrophy. J Neurol Sci 1979; 42: 337–347

63 Miller SE, Roses AD, Appel SH. Scanning electron microscopic studies in muscular dystrophies. Arch Neurol 1976; 33: 172–174

64 Lucy JA. Is there a membrane defect in muscle and other cells? Br Med Bull 1980; 36: 187–192

65 Shivers RR, Martin K, Atkinson BG. Detection of carriers of human DMD by freeze fracture analysis or erythrocyte plasmalemma intramembrane particles. Am J Clin Path 1986; 85: 131–134

66 Dubowitz V. Myopathic changes in muscular dystrophy carriers. Proc Roy Soc Med 1963; 56: 810–812

67 Emery AEH. Muscle histology in carriers of DMD. J Med Genet 1965; 2: 1–7

68 Smith HL, Amock LD, Johnson WW. Detection of subclinical and carrier states in DMD. J Paediatr 1966; 69: 67–79

69 Schiffer D, Bertolotto A, de Marchi et al. Epidemiology of Duchenne muscular dystrophy in the province of Turin. Ital J Neurol Sci 1981; 2: 81–84

70 Fisher ER, Wissinger A, Gometh JA, Danowski TS. Ultrastructural changes in skeletal muscle of muscular dystrophy carriers. Arch Pathol 1972; 94: 456–460

71 Afifi AK, Bergman R, Zellweger H. A possible role for electron microscopy in detection of carriers of Duchenne type muscular dystrophy. J Neurol Neurosurg Psychiatr 1973; 36: 643–650

72 Roy S, Dubowitz V. Carrier detection in DMD—a comparative study of electron microscopy, light microscopy and serum enzymes. J Neurol Sci 1970; 11: 65–79

73 Morris CJ, Raybould JA. Histochemically demonstrable fibre abnormalities in normal skeletal muscle and in muscles from carriers of DMD. J Neurol Neurosurg Psychiatr 1971; 34: 348–352

74 Maunder-Sewry CA, Dubowitz V. Needle muscle biopsy for carrier detection in Duchenne muscle dystrophy part 1. Light miscroscopy—histology, histochemistry and quantitation. J Neurol Sci 1981; 49: 305–324

75 Rott H-D, Santellani M, Rodl W, Nebel G. DMD: carrier detection by ultrasound and computerized tomography. Lancet 1983; II: 1199–1200

76 Rott H-D, Santellani M, Briemesser FH. DMD: carrier detection by ultrasound and computer tomography. Lancet 1984; I: 111

77 Stern LM, Caudrey DJ, Clark MS, Perrett LV, Boldt DW. Carrier detection in Duchenne muscular dystrophy using computed tomography. Clin Genet 1985; 27: 392–397

78 Castro-Gago M, Alonso A, Novo I, Fuster M. Carrier detection of DMD by computerized tomography. Lancet 1986; I: 1039

79 Schapira F, Dreyfus JC, Schapira G, Démos J. Etude de l'aldolase et de la créatine kinase du sérum chez les mères de myopathes. Rev Fr Etudes Clin Biol 1960; 5: 990–994

80 Thompson MW, Murphy EG, McAlpine PJ. An assessment of CK test in the detection of carriers of DMD. J Paediatr 1967; 71: 82–93

81 Walton JN. Carrier detection in X-linked muscular dystrophy. J Genet Hum 1969; 17: 497–501

82 Gruemer HD, Miller WG, Chinchilli VM et al. Prediction of carrier status in Duchenne dystrophy by creatine kinase measurement. Am J Clin Pathol 1985; 84: 655–658

83 Dubowitz V. Carrier detection and genetic counselling in Duchenne dystrophy. Dev Med Child Neurol 1975; 17: 352–368

84 Moss DW, Whitaker KB, Parmar C et al. Activity of CK in sera from healthy women carriers of DMD and cord blood, determined by the 'European' recommended method with NAC-EDTA activation. Clin Chim Acta 1981; 116: 209–216

85 Tippett PA, Dennis NR, Machin D, Price CP, Clayton DE. CK activity in the detection of carriers of DMD: comparison of two methods. Clin Chem Acta 1982; 121: 345–359

86 Penrose LS, Smith GF. (1966) Down's Anomaly. London: Churchill, 1966: 106–107

87 Dennis NR, Carter CO. Use of overlapping normal distributions in genetic counselling. J Med Genet 1978; 15: 106–108

88 Sibert JR, Harper PS, Thompson RJ. Carrier detection in DMD—evidence from a study of obligate carriers and mothers of isolated cases. In: Lunt GG, Marchbanks RM eds. The Biochemistry of Myasthenia Gravis and muscular dystrophy. London: Academic Press, 1978: 239–243

89 Emery AEH, Morton R. Genetic counselling in lethal X-linked disorders. Acta Genet (Basel) 1968; 18: 534–542

90 Baraitser M. Oxford Monographs in Medical Genetics: The Genetics of Neurological Disorders, revised ed. Oxford: Oxford University Press, 1985: 252–254

91 Bundey S, Crowley JM, Edwards JH, Westhead RA. Serum CK levels in pubertal, mature, pregnant and postmenopausal women. J Med Genet 1979; 16: 117–121

92 Nicholson GA, Gardner-Medwin D, Pennington RJT, Walton JN. Carrier detection in DMD: an assessment of the effect of age on detection rate with serum CPK activity. Lancet 1979; I: 692–694

93 Moser H, Vogt J. Follow-up study of serum creatine kinase in carriers of Duchenne muscular dystrophy. Lancet 1974; II: 661–662

94 Perry TB, Clarke-Fraser F. Variability of serum CPK activity in normal women and carriers of the gene for DMD. Neurology 1973; 23: 1316–1323

95 Dreifuss JC, Schapira F, Demos J, Rosa R, Schapira G. The value of serum enzyme determinations in the identification of dystrophic carriers. Ann NY Acad Sci 1966; 13: 304–314

96 Hausmanowa-Petrusewicz I, Prot J, Niebroj-Dobosz I, Hetnarska L, Emeryk B, Wasowicz B, Askanas W, Slucka C. Studies of healthy relative of patients with Duchenne muscular dystrophy. J Neurol Sci 1986; 7: 465–480

97 Munsat TL, Baloh R, Pearson CM, Fowler W. Serum enzyme alterations in neuromuscular disorders. J Am Med Assoc 1973; 226: 1536–1543

98 Zatz M, Frota-Pessoa O, Levy JA, Peres CA. Creatine phosphokinase (CPK) activity in relatives of patients with X-linked muscular dystrophies: a Brazilian study. J Genet Hum 1976; 24: 153–168

99 Gale AN, Murphy EA. The use of serum CPK in genetic counselling for Duchenne muscular dystrophy. II. Review of methods of assay and factors which may be relevant in the interpretation of serum CPK activity. J Chronic Dis 1979; 32: 639–651

100 Gaines RF, Pueschel SM, Sassaman EA, Driscoll JL. Effect of exercise on serum creatine kinase in carriers of Duchenne muscular dystrophy. J Med Genet 1982; 19: 4–7

101 Hudgson P, Gardner-Medwin D, Pennington RJ, Walton JN. Studies of the carrier state in the Duchenne type of muscular dystrophy—Part I (Effect of exercise on serum creatine kinase activity). J Neurol Neurosurg Psychiatr 1967; 30: 416–419

102 Hughes RC, Park DC, Parsons ME, O'Brien MD. Serum CK studies in the detection of carriers of Duchenne dystrophy. J Neurol Neurosurg Psychiatr 1971; 34: 527–530
103 Blyth H, Hughes BP. Pregnancy and serum CPK levels in potential carriers of severe X-linked muscular dystrophy. Lancet 1971; I: 855–856
104 Emery AEH, King B. Pregnancy and serum CK levels in potential carriers of Duchenne X-linked muscular dystrophy. Lancet 1971; I: 1013
105 King B, Spikesman A, Emery AEH. The effect of pregnancy on serum level of CK. Clin Chem Acta 1972; 36: 267–269
106 Bundey S. Calculation of genetic risks in Duchenne muscular dystrophy by geneticists in the United Kingdom. J Med Genet 1978; 15: 249–253
107 Emery AEH, Clack ER, Simon S, Taylor JT. Detection of carriers of benign X-linked muscular dystrophy. Br Med J 1967; 4: 522–523
108 Skinner R, Emery AEH, Anderson AJB, Foxall C. The detection of carriers of benign (Becker type) X-linked muscular dystrophy. J Med Genet 1975; 12: 131–134
109 Grimm T. Genetic counselling in Becker type X-linked muscular dystrophy. II. Practical considerations. Am J Med Genet 1984; 18: 719–723
110 Kingston HM, Sarfarazi M, Newcombe RG, Willis N, Harper PS. Carrier detection in BMD using CK estimation and DNA analysis. Clin Genet 1985; 27: 383–391
111 Hodgson SV. Genetic Studies in Duchenne Muscular Dystrophy. Thesis submitted for the degree of Doctor of Medicine, University of Oxford, 1987
112 Harper PS, O'Brien T, Murray JM, Davies KE, Pearson P, Williamson R. The use of linked DNA polymorphisms for genotype prediction in families with Duchenne muscular dystrophy. J Med Genet 1983; 20: 252–254
113 Pembrey ME, Davies KE, Winter RM et al. Clinical use of DNA markers linked to the gene for DMD. Arch Dis Child 1984; 59: 208–216
114 Hodgson AV, Walker A, Cole C et al. The application of linkage analysis to genetic counselling in families with Duchenne or Becker muscular dystrophy. J Med Genet 1987; 24: 152–159
115 Goodship J, Malcolm S, Robertson ME, Pembrey ME. Service experience using DNA analysis for genetic prediction in Duchenne muscular dystrophy. J Med Genet 1988; 25: 14–19
116 Fischbeck KH, Ritter AW, Tirschwell DL et al. Recombination with pERT 87 (DXS164) in families with X-linked muscular dystrophy. Lancet 1986; II: 104
117 Walker A, Hart K, Cole C et al. Linkage studies in Duchenne and Becker muscular dystrophies. J Med Genet 1986; 23: 538–547
118 Monaco AP, Bertelson CJ, Middlesworth W et al. Detection of deletions spanning the DMD locus using a tightly linked DNA segment. Nature 1985; 316: 842–845
119 Ray PN, Belfall B, Duff C et al. Cloning of the breakpoint of an X;21 translocation associated with Duchenne muscular dystrophy. Nature 1985; 318: 672–675
120 Williams H, Sarfarazi M, Brown C, Thomas N, Harper PS. The use of flanking markers in prediction for DMD. Arch Dis Child 1986; 61: 218–222
121 Lathrop GM, Lalouel JM, Julier C, Ott J. Strategies for multilocus linkage analysis in humans. Proc Natl Acad Sci USA 1984; 81: 3443–3446
122 Sarfarazi M, Williams H. (1986) A computer programme for estimation of genetic risk in X-linked disorders, combining pedigree and DNA probe data with other conditional information. J Med Genet 1986; 23: 40–45
123 Jeffreys AJ, Wilson V, Thein SL. Hypervariable minisatellite regions in human DNA. Nature 1985; 316: 67–73
124 Koh J, Bartlett RJ, Pericak-Vance MA et al. Inherited deletion at Duchenne dystrophy locus in normal male. Lancet 1987; II: 1155

125 Hart K, Abbs S, Bobrow M. Pathogenic and non-pathogenic deletions in two families with Duchenne muscular dystrophy. Submitted for publication, 1989

126 Hart K, Cole C, Walker A et al. The screening of Duchenne muscular dystrophy patients for sub-microscopic deletions. J Med Genet 1986; 23: 510–520

127 Kunkel LM et al. Analysis of deletions in DNA from patients with Becker and Duchenne muscular dystrophy. Nature 1986; 322: 73–77

128 Worton RG. Molecular analysis of Duchenne and Becker muscular dystrophy. BioEssays 1987; 7: 57–62

129 Bartlett RJ, Pericak-Vance MA, Koh J et al. Duchenne muscular dystrophy: high frequency of deletions. Neurology 1988; 38: 1–4

130 Lindlöf M, Kaariainen H, van Ommen G-JB, de la Chapelle A. Microdeletions in patients with X-linked muscular dystrophy: molecular-clinical correlations. Clin Genet 1988; 33: 131–139

131 Forrest SM, Smith TJ, Cross GS. Effective strategy for prenatal diagnosis of Duchenne and Becker muscular dystrophy. Lancet 1987; II: 1294–1297

132 den Dunnen JT, Bakker E, Klein Breteler EG, Pearson PL, van Ommen G-JB. Direct detection of more than 50% of the DMD mutations by field inversion gels. Nature 1987; 329: 640–642

133 Hu X, Burghes AHM, Ray PN, Thompson MW, Murphy EG, Worton RG. Partial gene duplication in Duchenne and Becker muscular dystrophies. J Med Genet 1988; 25: 369–376

134 Bertelson CJ, Bartley JA, Monaco AP, Colletti-Feener C, Fischbeck K, Kunkel LM. Localisation of Xp21 meiotic exchange points in DMD families. J Med Genet 1986; 23: 531–537

135 Bakker E, van Broeckhoven C, Bonten EJ et al. Germline mosaicism and DMD mutations. Nature 1987; 329: 554–556

136 Lanman JT, Pericak-Vance MA, Bartlett RJ et al. Familial inheritance of a DXS164 deletion mutation from a heterozygous female. Am J Hum Genet 1987; 41: 138–144

137 Wood S, McGillivray BC. Germinal mosaicism in DMD. Hum Genet 1988; 78: 282–284

138 Darras BT, Francke U. A partial deletion of the MD gene transmitted twice by an unaffected male. Nature 1987; 329: 556–558

139 Cole CG, Walker A, Coyne A et al. Prenatal testing for Duchenne and Becker muscular dystrophy. Lancet 1988; I: 262–266

140 Toop J, Emery AEH. Muscle histology in fetuses at risk for Duchenne muscular dystrophy. Clin Genet 1974; 5: 230–233

141 Emery AEH. Pathogenesis of Human Muscular Dystrophy. In: Rowlands LP ed. International Congress series no. 404. Amsterdam: Excerpta Medica, 1977: 42–52

142 Winn KJ, Heller RH. Pathologic diagnosis of DMD in an aborted fetus. Clin Genet 1978; 13: 335–338

143 Mahoney MJ, Haseltine FP, Banker BQ, Hobbins JC, Caskey CT, Golbus MS. Prenatal diagnosis of DMD. New Engl J Med 1977; 297: 968–973

144 Dubowitz V, Rodeck CH, Campbell S, Singer JD, Scheuerbrandt G, Moss DW. Prenatal diagnosis in DMD. Salvage of a normal male fetus. Lancet 1978; I: 90

145 Ionasescu V, Cancilla P. Fetal serum creatine phosphokinase not a valid predictor of Duchenne muscular dystrophy. Lancet 1978; II: 1251

146 Emery AEH, Dubowitz V, Rocker I, Donnai D, Harris R, Donnai P. Antenatal diagnosis of Duchenne muscular dystrophy. Lancet 1979; I: 847–849

147 Edwards RJ, Rodeck CH, Watts DC. Plasma CK and myoglobin levels, before and after abortion in human fetuses at risk for DMD. Am J Med Genet 1983; 15: 475–482

148 Watts DC. Problems in the prenatal diagnosis of DMD. Biochem Soc Trans 1984; 12: 366–368

149 Török O, Szabo M, Veress L, Papp Z. Plasma/serum myoglobin in prenatal diagnosis of DMD. Am J Med Genet 1986; 25: 237–239

150 Donnai D, Charles IV, Harris R. Attitudes of patients after 'genetic' termination of pregnancy. Br Med J 1981; 282: 621–622

151 Williams H, Brown CS, Thomas NST et al. First trimester fetal sexing in pregnancy at risk for Duchenne muscular dystrophy. Lancet 1983; II: 568–569

152 Gustavii B, Chester MA, Edvall H et al. First-trimester diagnosis on chorionic villi obtained by direct vision techniques. Hum Genet 1984; 65: 373–376

153 Crane JP, Beaver HA, Cheung SW. First trimester chorionic villus sampling versus mid-trimester genetic amniocentesis—preliminary results of a controlled prospective trial. Prenatal Diagnosis 1988; 8: 355–366

154 Darras BT, Koenig M, Kunkel LM, Francke U. Direct method for prenatal diagnosis and carrier detection of D/BMD using the entire dystrophin gene. Am J Med Genet 1988; 29: 713–726

155 Rudolph N, Cross R. CPK activation in serum of newborn infants as an indicator of fetal trauma during birth. Pediatrics 1966; 38: 1039–1046

156 Griffiths PD. The activation of ATP: creatine PK (E.C.2.7.3.2.) in umbilical cord blood. Clin Chim Acta 1968; 20: 465–472

157 Bodensteiner JS, Zellweger H. CPK in normal neonates and young infants. J Lab Clin Med 1971; 77: 853–857

158 Zellweger H, Antonik A. Newborn screening for DMD. Paediatrics 1975; 55: 30–34

159 Gilboa N, Swanson J. Serum CPK in normal newborns. Arch Dis Child 1976; 51: 283–285

160 Thompson MW. Genetic management of pregnancies of carriers and possible carriers of Duchenne muscular dystrophy. In: Serratrice G et al. eds. Neuromuscular Disease. New York: Raven Press, 1984: 21–23

161 Zatz M. Duchenne muscular dystrophy: observations on the effects of genetic counselling in Brazil. In: Serratrice G et al. eds. Neuromuscular Disease. New York: Raven Press, 1984: 17–19

162 Chapple JC, Dale R, Evans BG. The new genetics: will it pay its way? Lancet 1987; I: 1189–1192

163 Lench N, Stanier P, Williamson R. Simple non-invasive method to obtain DNA for gene analysis. Lancet 1988; I: 1356–1358

British Medical Bulletin (1989) Vol. 45, No. 3, pp. 745–759
© The British Council 1989

Myotonic dystrophy: Developments in molecular genetics

D J Shaw
P S Harper
Institute of Medical Genetics, University of Wales College of Medicine, Cardiff, Wales, UK

Myotonic dystrophy, the commonest muscular dystrophy of adult life and the most variable of all muscular dystrophies follows autosomal dominant inheritance and is determined by a genetic locus on chromosome 19. The development of DNA probes on this chromosome, some of them representing specific genes, has provided a series of closely linked markers for myotonic dystrophy, which together with the construction of a series of somatic cell hybrid lines has allowed a detailed localization on the long arm of the chromosome. Existing markers are already able to provide accurate tests of genetic prediction, including prenatal diagnosis. Analysis of the relevant region of 19q by pulsed field gel electrophoresis and other molecular techniques is now in progress to identify the myotonic dystrophy gene itself.

Myotonic dystrophy, the commonest muscular dystrophy of adult life, is also the most variable clinically in both severity and mode of presentation. This variability produces difficulties not only in diagnosis but in genetic counselling and presymptomatic detection. Until now, no primary underlying defect or constant biochemical abnormality has been identified that can be of help; thus the accurate localization and isolation of the gene is of particular importance for genetic prediction as well as for an understanding of the molecular pathology of the disorder. Since most recent research of importance on myotonic dystrophy has been in the field of molecular genetics, the present review will concentrate on

0007–1420/89/0045–0745/$10.00

this aspect of the disorder, dealing briefly first with those clinical aspects that relate most closely to the genetic basis; fuller clinical reviews are given elsewhere.[1,2]

Neuromuscular features

Myotonic dystrophy, in common with other forms of muscular dystrophy, results in progressive muscle weakness and wasting, whose distribution (Fig. 1) is characteristic. While there may be a superficial resemblance to the pattern seen in facioscapulohumeral dystrophy, and to that of myasthenia gravis the combination of facial, jaw and anterior neck muscle involvement with distal limb weakness allows a ready distinction from most other neuromuscular disorders, notably from Becker and Duchenne dystrophies which involve predominantly the pelvic girdle and proximal limb muscles. However, many patients with myotonic dystrophy show minimal muscle weakness, while some of those with overt weakness on testing may not complain of it.

Myotonia, the most characteristic feature of myotonic dystrophy, is interpreted by most patients as muscle stiffness and can be seen in the hands as delayed relaxation of active grip, as well as on direct percussion of the thenar muscles. Myotonia may be more widespread, affecting the periorbital and jaw muscles, and is often aggravated by cold. As with the weakness, many patients minimize their myotonia, some even considering it a normal feature. Thus, while the clinical diagnosis of myotonic dystrophy presents no problems in the patient with definite myotonia and weakness of characteristic distribution, it is often far from simple in individuals who present with non-neuromuscular symptoms, in those with mild disease, or in relatives at risk, where myotonia and weakness may be minimal. Myotonia alone may not be sufficient for the diagnosis, since a group of rare non-progressive disorders characterized by myotonia exists (Table 1), from which distinction may be difficult in an isolated case where myotonia is the predominant feature.

Systemic involvement

Myotonic dystrophy, much more than any other muscular dystrophy, is a multisystem disorder (Table 2). Not only are smooth and cardiac muscle commonly involved, but also the eye, central nervous system and endocrine organs. Cataract is a frequent

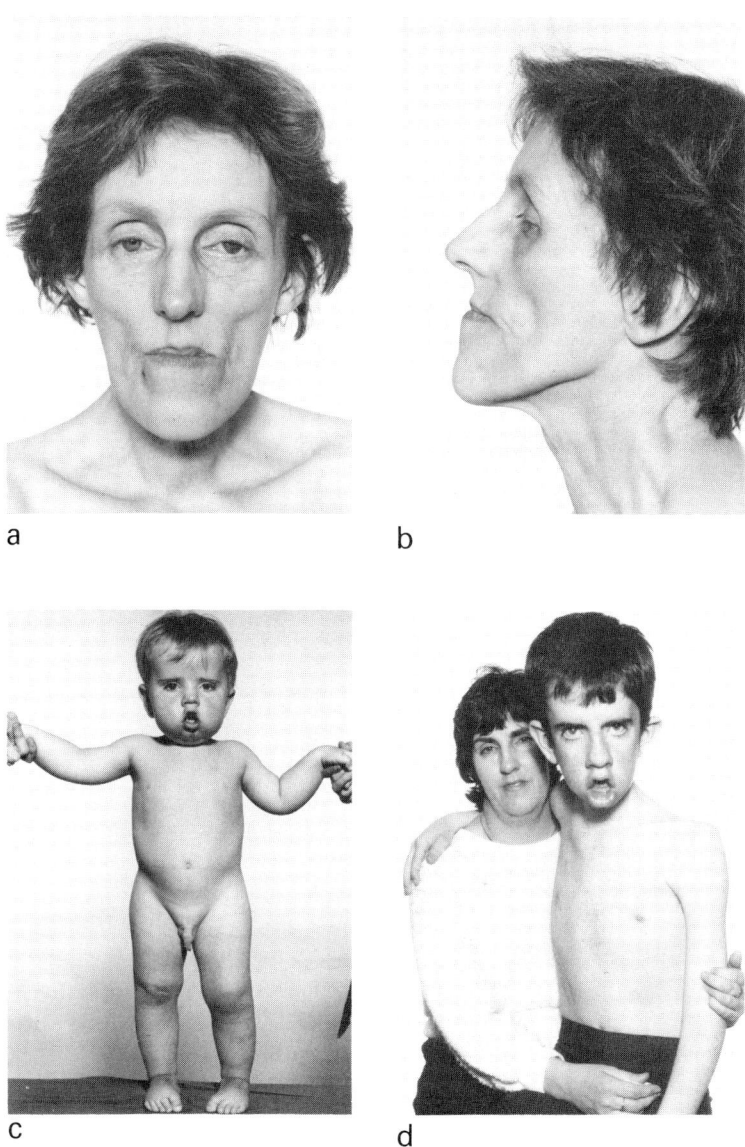

Fig. 1 Muscle involvement in myotonic dystrophy.
ab. Affected adult to show weakness and wasting of facial jaw and neck muscles
c. Affected child with congenital myotonic dystrophy, presenting with hypotonia, motor delay and facial diplegia
d. Same child age 14 years, with mildly affected mother

Table 1 The nonprogressive myotonias

Disorder	Clinical Features	Inheritance
Myotonia congenita (Thomsen's disease)	Lifelong myotonia	Autosomal dominant
Recessive generalised myotonia	Myotonia often progressive in early years; muscle hypertrophy marked	Autosomal recessive
Paramyotonia congenita	Prolonged cold induced contractures as well as myotonia	Autosomal dominant
Myotonic periodic paralysis (adynamia hereditaria)	Intermittent flaccid weakness; myotonia mild	Autosomal dominant
Chondrodystrophic myotonia (Schwarz Jampel syndrome)	Severe progressive bone dysplasia in addition to myotonia	Autosomal recessive

Table 2 Involvement of various organ systems in myotonic dystrophy

System	Principal features
Muscle	Myotonia; weakness and wasting in typical distribution
Central nervous system	Somnolence, apathy (adults); mental retardation behavioural problems (children)
Heart	Conduction defects, cardiomyopathy
Gastrointestinal tract	Swallowing difficulties, colonic symptoms (rarely dilatation)
Endocrine	Testicular tubal atrophy, diabetes
Eye	Cataract, retinal degeneration, reduced intra-ocular pressure

mode of presentation in middle-aged or elderly patients; the characteristic multi-coloured lens opacities are highly specific in their early stages, and form one of the best documented presymptomatic tests for gene carriers.[3] Involvement of the central nervous system is especially prominent in congenital cases (see below); learning difficulties and behavioural problems are frequent in affected children, even when intelligence is normal, while in later life apathy and excessive somnolence can be conspicuous along with numerous minor psychometric and personality abnormalities.[4] Endocrine involvement is associated with testicular atrophy and loss of tubular function in males[5] and with a high fetal loss in females,[6] fertility being significantly reduced in both sexes.

Congenital myotonic dystrophy

In addition to the generally variable age at onset of myotonic dystrophy, which can frequently be detected in childhood even in the absence of symptoms, a specific form of the disease may occur in the offspring of affected mothers.[7,8] Here the disorder presents at birth, often with evidence of prenatal onset. Prominent features include respiratory inadequacy, with hypoplasia of respiratory muscles, general hypotonia and motor delay, while in survivors mental retardation of varying degree is usual. Polyhydramnios, talipes and reduced fetal movements indicate the prenatal presence of neuromuscular failure of swallowing and movement. The relevant features from the genetic viewpoint are that this form is not confined to a specific subgroup of families, that it is exclusively maternally transmitted, and that surviving children later develop progressive features of more typical myotonic dystrophy.[9] Those sibs who are apparently unaffected remain so, suggesting that only those inheriting the gene can develop the congenital form. Hypotheses relating to this unusual situation include a transmissible maternal factor,[10] mitochondrial inheritance and genomic imprinting, but no clear evidence yet exists for any of these explanations.

It should be noted that departure from the expected pattern of autosomal dominant inheritance may also exist at the other extreme of life, where study of the earliest available generation may show minimal features of the disorder in a grandparent (most commonly the grandfather), making it almost impossible to identify a new mutation with confidence. This remarkable variation between the generations, recognized for many years under the name of 'anticipation', has for long been held to be artefact of ascertainment bias,[11] but may prove, as in other disorders such as fragile X syndrome, to have a biological basis.

Experimental approaches to the primary defect

It now seems likely that the protein abnormality in myotonic dystrophy will be recognized by the 'reverse genetics' approach, through development of the molecular genetics research described below. When this point is reached it will become relevant to correlate this work with what is already known from other approaches. While details of this have been fully summarized elsewhere[1,2], the three principal areas of work are outlined here:

    1. Electrophysiological studies of myotonia in affected muscle

and in animal models.[12,13] These have clearly shown that myotonia may result from different mechanisms which affect the excitability of the muscle membrane. The reduced resting membrane potential seen in myotonic dystrophy is distinct from the abnormalities seen in non-progressive myotonic disorders of man and other species. Isolation of the specific proteins and genes involved in ion transport is beginning to provide useful candidate genes for the latter group, which appear to be non-allelic to myotonic dystrophy.

2. Biochemical and biophysical studies of cell membranes other than muscle have detected abnormalities which have proved controversial and difficult to reproduce.[14-16] A generalized membrane defect would certainly fit with the widespread nature of the disorder. The localization of the Na^+/K^+ ATPase gene on chromosome 19 raised the possibility that this gene might be responsible, but this has now been excluded.[17]

3. Muscle specific proteins. While muscle histology and ultrastructure in myotonic dystrophy show characteristic abnormalities that are useful diagnostically, no defect in any of the principal contractile or other muscle proteins has been found.[18] The recognition in Duchenne muscular dystrophy that dystrophin is indeed a cytoskeletal protein though one of exceedingly low abundance in muscle (see Kunkel et al., this volume), raises the possibility that comparable proteins may be responsible for other muscular dystrophies, including myotonic dystrophy.

MOLECULAR GENETICS

Background

Despite much effort having been expended on investigation of the underlying defect using the approaches outlined above, no clear picture has emerged. Thus the approach of 'reverse genetics', where the isolation of a disease gene is based on its genetic map location without prior knowledge of its base sequence or protein product is particularly attractive. This approach has already led to the isolation of the Duchenne muscular dystrophy gene, discussed elsewhere in this volume.

The studies of Renwick et al.[19] and Harper et al.[20] confirmed the earlier report of Mohr[21] that DM might be genetically linked to the Lutheran blood group and secretor loci. Eiberg et al.[22] subsequently showed that complement C3 was part of the same linkage group, which was then assigned to chromosome 19 by somatic-cell hybrid mapping of the C3 locus.[23]

Since that time, many cloned genes and random DNA sequences from chromosome 19 have been isolated.[24] Those that detect RFLPs have been used extensively in family linkage studies, and have been localized to specific subchromosomal regions using chromosome 19 translocations in somatic cell hybrids.

The first report of a DNA marker closely linked to DM was that of Shaw et al.[25] who showed that the apolipoprotein C2 locus (APOC2) was about 4cM from DM. APOC2 has since proved to be a useful marker for the prenatal diagnosis of DM.[26] The marker has been studied by a number of groups worldwide and pooled data indicate a recombination fraction of 0.02, with a lod score approaching 100, and no evidence for genetic heterogeneity.

Several studies have now shown that the DM locus is on the proximal long arm of chromosome 19. Friedrich et al.[27] used multipoint linkage analysis of DM, complement C3 and a cytogenetic centromere polymorphism, to demonstrate the order C3–cen–DM. (C3 had previously been localized to the short arm of 19 by Brook et al.[28]). The APOC2 locus was assigned to

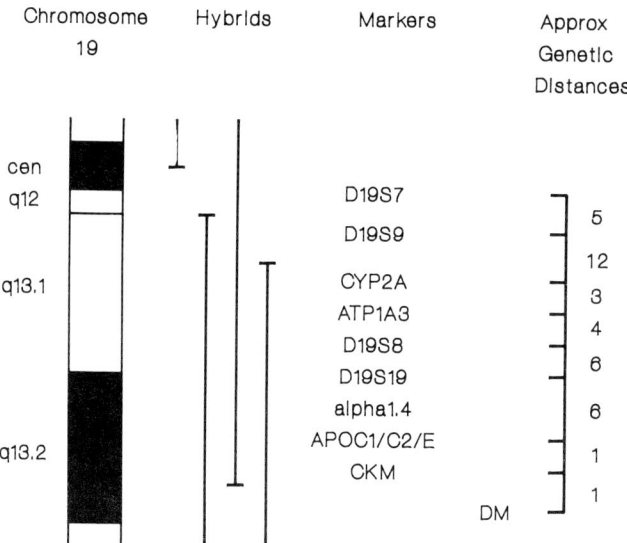

Fig. 2 Principal DNA markers on long arm of chromosome 19 in relation to localization of the myotonic dystrophy gene.

proximal 19q by somatic cell hybrid studies and by *in situ* hybridization.[29,30]

The locus D19S19 (probe LDR152) also shows close linkage to DM and was assigned to proximal 19q by Bartlett et al.[31]

Accurate genetic mapping of the DM locus

More recent work has focused on the isolation and characterization of new DNA markers from the DM region, for fine mapping of the locus. To this end somatic cell hybrids have been constructed which contain the relevant part of chromosome 19, and little other human material, on a rodent background.[29,32,33] Random DNA sequences as well as cloned genes from other sources have been studied in terms of their genetic linkage relationships and subchromosomal localizations determined from independent somatic cell hybrid mapping panels. The markers fall into 3 groups, as follows:

1. A centromeric group of markers (19cen–19q12). This includes D19S7 (probe 4.1[30]), and also D19S13.[34]

2. A group of markers localized to 19q12–q13.1, by the use of somatic cell hybrids with defined breakpoints on chromosome 19[34,35] or with fragmented derivatives (Brook et al., unpublished data). This group includes D19S9 (probe p1J2[30]; the genes for cytochrome p450 (CYP2A), peptidase D (PEPD), glucose phosphate isomerase (GPI), transforming growth factor beta (TGFβ), myelin associated glycoprotein (MAG) and a cell-surface antigen (MSK37) of unknown function; and a number of random DNA markers.

3. A group located at 19q13.1–q13.2, that contains all of the markers most closely linked to DM.

(i) The sodium potassium ATPase alpha 3 subunit (ATP1A3). This gene is of particular relevance to DM because of earlier biochemical studies that suggested it might have a role in the pathology of the disease.[36] In a recent study[17] we showed that ATP1A3 is linked to DM at a distance of 9cM, with a number of clear recombination events that rule out the possibility of ATP1A3 being the disease gene itself.

(ii) The DNA markers D19S8, D19S15 and D19S16. These loci are all linked to DM at distances of 3–6cM.[37] These probes are of potential value in diagnosis of DM when other, more closely-linked markers (see below) are uninformative.

(iii) The DNA locus D19S19 (probe LDR152). The original

study[31] showed this marker to be closely linked to DM, with no definite recombination events; in subsequent worldwide studies recombinants have been detected (unpublished data). Based on these a realistic estimate of the D19S19–DM distance is about 4cM.

(iv) The apolipoprotein gene cluster. Since the original linkage studies on APOC2, and assignment of APOC2, apolipoprotein C1 (APOC1) and apolipoprotein E (APOE) to the region 19q12–q13.2, it has been shown by pulsed-field gel electrophoresis that all three APO genes are within 50 kb of each other.[38] The linkage of APOC2 to DM has been described earlier, and other studies have shown that APOE and APOC1 are also linked to DM.[39] Despite the very close physical linkage of the 3 APO genes, several instances of recombination between them suggest the presence of a 'hot-spot' and indicate that combining RFLPs at all 3 genes to form haplotypes may not be justified.

(v) A chronic lymphocytic leukaemia breakpoint region. McKeithan et al.[40] reported the cloning of a 14, 19 translocation breakpoint, involving the IGH locus on 14 and q13.1 region of 19, from a patient with chronic lymphocytic leukaemia (CLL). A probe from this locus (alpha 1.4P) detects an RFLP which is closely linked to DM.[41] We have shown by pulsed-field gel electrophoresis that the CLL breakpoint is about 200 kb proximal to the APO gene cluster.[42]

(vi) The creatine kinase muscle isoform gene (CKMM). Experiments using somatic cell hybrids localized this gene to the APOC2 region of 19q.[33] Subsequent studies using a probe from the 3' end of the gene,[34] have shown that this gene is very closely linked to DM, at a distance estimated from our own studies at less than 2cM (unpublished results).

(vii) The poliovirus receptor gene (PVS). Although it has not yet been cloned it has been possible to localize this gene to the DM region using somatic cell hybrids, because its presence in a hybrid cell line makes the cells susceptible to killing by poliovirus.[35] Once the gene has been isolated it should serve as another useful linkage marker for this region.

Most of the markers listed above have been used in genetic linkage studies, with a view to determining their order relative to each other and to DM. Nakamura et al.[43] published a genetic map of chromosome 19, part of which involves some of the above markers (although not the DM gene). This study confirmed the marker order obtained using somatic cell hybrids, as follows:

cen–D19S7–D19S9–APOC2–qter

Our own results, from both DM and reference (CEPH) families, have enabled an overall maximum likelihood order of the loci on proximal 19q to be deduced. The map is derived from both computer analysis of linkage data (2 point and multipoint) and analysis of multiply-informative meioses where recombination events have occurred:

cen–D19S7–D19S9–CYP2A–ATP1A3–D19S8–D19S19–CLL
breakpoint–APOC1/C2/E–(DM, CKMM)–qter

Other markers have still to be evaluated as flanking the DM locus. These include D19S22 which has been localized 11cM distal to APOC2,[43] and protein kinase C gamma (PKCG) which is in the region 19q13.3–q13.4[44] (Brook et al., unpublished data).

Application in prediction

Myotonic dystrophy was one of the first disorders where genetic linkage (in this case the secretor locus) was applied in prediction in prenatal diagnosis.[45,46] Looseness of linkage and inability to distinguish the heterozygous state for the marker precluded more than occasional use, so that systematic application had to await closely linked DNA markers. The APOC2 gene probe, with recombination frequency around 2% and showing a number of polymorphisms, has proved useful in both prenatal and presymptomatic prediction;[26] the recognition that APOE and APOC1 are adjacent to APOC2 and the mapping of further loci such as CKMM even closer to myotonic dystrophy now allows over 90% of families to be informative for prediction. The use of the polymerase chain reaction (PCR) is already feasible for some of these markers for which sequence data are available. Our own experience has been that prenatal diagnosis is requested principally by those families where a congenitally affected child has already been born. It is particularly important that DNA should be isolated from such infants, who may well die in the neonatal period.

PROGRESS TOWARDS ISOLATION OF THE DM GENE

As described above, the DM gene has been localized to a small region of 19q. Based on the generally accepted estimate of the ratio of physical to genetic distance in the human, we estimate that DM is located about 2 megabases (2Mb) distal to APOC2, and probably closer to CKMM. Recently Stallings et al.[33] have described a hybrid cell line that contains a fragment of chromosome 19 as its only human material. This fragment contains the markers closest to DM and hence probably also the DM gene itself, and is being used as starting material for isolation of the gene. Although it is not difficult (in principle) to obtain large numbers of candidate genes from a hybrid cell line, the real problem lies in proving which of these is responsible for myotonic dystrophy. Some possible approaches are considered below.

(i) Pulsed-field gel (PFG) mapping is a technique that can be readily applied to the physical characterisation of megabase regions of human DNA. Existing probes can be linked together and ordered, and new ones generated from specific regions of the map by cutting out DNA fragments from pulsed-field gels and making libraries from them. If DM is caused by a deletion of DNA then this could be detected by PFG, using a relatively small number of probes and DNA samples from a selection of patients. With hindsight it is evident that such an approach would have worked in the case of Duchenne muscular dystrophy.

(ii) Linkage disequilibrium. Conventional linkage analysis relies on the observation of recombination events between the loci under study to deduce their relative order. In the case of extremely closely-linked loci, recombination will not be observed at a significant frequency, and the loci will tend to show linkage disequilibrium (or allelic association). This means that the assortment of alleles for each locus is not random in the population. Diseases such as myotonic dystrophy and cystic fibrosis (CF) have a very low rate of new mutation, and in the case of CF it has already been shown that various haplotypes of the closely-linked marker loci are strongly associated with the CF mutation.[47] None of the DM linked probes studied to date shows general linkage disequilibrium with DM, although disequilibrium has been demonstrated in specific, geographically isolated populations.[41] It is likely that markers less than 1cM from the DM gene will begin to show disequilibrium, and that (with some exceptions) the closer they are, the stronger the effect will be. Over these very short (in genetic terms) distances, disequilibrium studies should prove more useful than pedigree recombinational analysis.

(iii) Gene isolation. Since most human DNA is non-coding it is necessary to have an efficient method for identifying the expressed sequences. Many genes possess at their 5′ ends short stretches of under-methylated, GC rich DNA called HTF islands. These were originally identified because the enzyme HpaII cuts them into 'Tiny Fragments' and they are often apparent in PFG studies because the enzymes used in such experiments are usually specific for GC rich, unmethylated target sites.[48] A second method, used successfully by Monaco et al.[49] in the isolation of the DMD gene, is to screen DNA clones for inter-species homology with Southern blots of DNAs from different animal species ('Zoo-blots'). A third procedure is to screen the clones against blots of mRNA from different tissues ('Northern blots') although this will not work if the gene is expressed very weakly or not at all in the tissues from which the RNA was isolated.

(iv) Testing candidate genes. It is quite easy to show that a candidate is not the DM gene, but how to prove conclusively that a candidate is the true DM gene is less obvious. Assuming that the gene is not exceptionally large, RFLPs detected by the gene probe should show no recombination with the DM phenotype, and probably also be in linkage disequilibrium with it. The actual molecular defect may be a deletion or insertion of DNA, which would be readily detectable; in the most difficult case it might be a single base change, which would necessitate sequencing of the candidate gene from normal and DM chromosomes (or other methods capable of detecting single base changes such as denaturing gradient gel electrophoresis[50]). The same gene sequenced from any two random chromosomes would be expected to show a number of differences that might have no effect on its function, so it would be important to establish that any putative DM mutation is in fact exclusively associated with the disease. A genuine candidate gene would also show a pattern of tissue-specific expression compatible with the pathology of the disease, although in the case of myotonic dystrophy it is not clear what this pattern should be. Ultimate characterization of a candidate might involve its expression in a transgenic mouse (this is theoretically possible because the DM mutation is dominant) and study of the effects it produces; there are unfortunately no natural animal models for the disease, in contrast to the non-progressive myotonias.

CONCLUSION

Genetic mapping using recombinant DNA probes has been highly successful in providing an accurate localization and the means of prenatal prediction for myotonic dystrophy, as indeed it has for many other genetic diseases. New methods are now being applied for the isolation of DM candidate genes from the relevant region of chromosome 19, and it is likely that the DM gene will be identified in the near future. This should provide the starting point for an understanding of the molecular pathology, and perhaps even treatment, of this relatively common inherited disease.

ACKNOWLEDGEMENTS

We should like to acknowledge the support of the Muscular Dystrophy Group of Great Britain, the Muscular Dystrophy Association of America, the Wolfson Foundation, the Wellcome Trust, and the Medical Research Council.

REFERENCES

1 Harper PS. Myotonic Dystrophy. Philadelphia: Saunders, 1979; 2nd edition 1989
2 Harper PS. Myotonic Disorders. In: Engel AG, Banker BQ, eds. Myology. New York: McGraw Hill, 1986: pp. 1267–1293
3 Harper PS. Presymptomatic detection and genetic counselling in myotonic dystrophy. Clin Genet 1973; 4: 134–140
4 Bird TD, Follett C, Grieg E. Cognitive and personality function in myotonic muscular dystrophy. 1. Cognitive function; 2. Personality profiles. J Neurol Neurosurg Psychiatry 1983; 46: 971–980
5 Harper PS, Penny R, Foley Jr T, Migeon CJ, Blizzard RM. Gonadal function in males with myotonic dystrophy. J Clin Endocrinol Metab 1972; 35: 852–856
6 O'Brien TA, Harper PS. Reproductive problems and neonatal loss in women with myotonic dystrophy. J Obstet Gynaecol 1984; 4: 170–173
7 Watters GV, Williams TW. Early onset myotonic dystrophy. Arch Neurol 1967; 17: 137–152
8 Dyken PR, Harper PS. Congenital dystrophia myotonica. Neurology 1973; 23: 465–473
9 Harper PS. Congenital myotonic dystrophy in Britain. II. Genetic basis. Arch Dis Child 1975; 50: 514–521
10 Harper PS, Dyken PR. Early onset dystrophia myotonica—Evidence supporting a maternal environmental factor. Lancet 1972; 2: 53–55
11 Penrose LS. The problem of anticipation in pedigrees of dystrophia myotonica. Ann Eugen (Lond) 1984; 14: 125–132
12 Rudel R. The Pathophysiologic basis of the myotonias and periodic paralyses. In: Engel AG, Banker BQ, eds. Myology. New York: McGraw Hill, 1986; pp. 1297–1312
13 Bretag AH. Muscle chloride channels. Physiol Rev 1987; 67: 618–724
14 Roses AD, Appel SH. Phosphorylation of component A of the human erythrocyte membrane in myotonic muscular dystrophy. J Membr Biol 1975; 20: 51–58
15 Butterfield DA, Chestnut DB, Appel SH, Roses AD. Spin label study of erythrocyte membrane fluidity in myotonic and Duchenne dystrophy and congenital myotonia. Nature 1976; 263: 159–161
16 Lucy JA. Is there a membrane defect in muscle and other cells? Br Med Bull 1980; 36: 187–192
17 Harley HG, Brook JD, Jackson C, et al. Localisation of human Na +, K + −

ATPase alpha subunit gene to chromosome 19q12–q13.2 and linkage to the myotonic dystrophy locus. Genomics 1988; 3: 380–384

18 Gergely J. Contractile proteins. In: Rowland LP, ed. Pathogenesis of Human Muscular Dystrophy. Excerpta Medica: Amsterdam, 1977

19 Renwick JH, Bundey SE, Ferguson-Smith MA, Izatt MM. Confirmation of the linkage of the loci for myotonic dystrophy and ABH secretion. J Med Genet 1971; 8: 407–416

20 Harper PS, Rivas ML, Hutchinson JR, McKusick VA, Bias WB, Dyken PR. Genetic linkage confirmed between the locus for myotonic dystrophy and the ABH secretion and Lutheran Blood group loci. Am J Genet 1972; 24: 310–316

21 Mohr J. A study of linkage in man. 1954. Copenhagen, Munksgaard

22 Eiberg H, Mohr J, Nielsen LS, Simonsen N. Genetics and linkage relationships of the C3 polymorphism: discovery of the C3:Se linkage and assignment of LES–C3–DM–Se–PEPD–LU synteny to chromosome 19. Clin Genet 1983; 24: 159–170

23 Whitehead AS, Solomon E, Chambers S, Bodmer WF, Povey S, Fey G. Assignment of the structural gene for the third component of human complement to chromosome 19. Proc Natl Acad Sci USA 1982; 79: 5021–5025

24 Shaw DJ, Eiberg H. Report of the Committee for chromosomes 17, 18, and 19. Cytogenet Cell Genet 1987; 46: 242–256

25 Shaw DJ, Meredith AL, Sarfarazi M et al. The apolipoprotein C2 gene: subchromosomal localisation and linkage to the myotonic dystrophy locus. Hum Genet 1985; 70: 271–273

26 Meredith AL, Huson SM, Lunt PW et al. Application of a closely linked polymorphism of restriction fragment length to counselling and prenatal testing in families with myotonic dystrophy. Br Med J 1986; 293: 1353–1356

27 Friedrich U, Brunner H, Smeets D, Lambermon E, Ropers HH. Three point linkage analysis employing C3 and 19cen markers assigns the myotonic dystrophy gene to 19q. Hum Genet 1987; 75: 291–293

28 Brook JD, Shaw DJ, Meredith AL, Bruns GAP, Harper PS. Localisation of genetic markers and orientation of the linkage group on chromosome 19. Hum Genet 1984; 68: 282–285

29 Hulsebos T, Wieringa B, Hochstenbach R et al. Toward early diagnosis of myotonic dystrophy: construction and characterisation of a somatic cell hybrid with a single human der (19) chromosome. Cytogenet Cell Genet 1986; 43: 47–56

30 Shaw DJ, Meredith AL, Sarfarazi M et al. Regional localisations and linkage relationships of seven RFLPs and myotonic dystrophy on chromosome 19. Hum Genet 1986; 74: 262–266

31 Bartlett RJ, Pericak-Vance MA, Yamaoka L. A new probe for the diagnosis of myotonic muscular dystrophy. Science 1987; 235: 1648–1650

32 Brook JD, Shaw DJ, Thomas NST, Meredith AL, Cowell J, Harper PS. Mapping genetic markers on human chromosome 19 using subchromosomal fragments in somatic cell hybrids. Cytogenet Cell Genet 1986; 41: 30–37

33 Stallings RL, Olson E, Strauss AW, Thompson LH, Bachinski LL, Siciliano MJ. Human creatine kinase genes on chromosomes 15 and 19, and proximity of the gene for the muscle form to the genes for apolipoprotein C2 and excision repair. Am J Hum Genet 1988; 43: 144–151

34 Schonk D, Coerwinkel-Driessen M, Van Dalen I, et al. Definition of subchromosomal intervals around the myotonic dystrophy region at 19q. Genomics 1988 (in press)

35 Brook JD, Skinner M, Roberts SH, Rettig WJ, Almond JW, Shaw DJ. Further mapping of markers around the centromere of human chromosome 19. Genomics 1987; 1: 320–328

36 Hull KL, Roses AD. Stoichiometry of sodium and potassium transport in

erythrocytes from patients with myotonic dystrophy. J Physiol 1976; 254: 169–181

37 Brunner H, Lambermon H, Hulsebos T et al. Multipoint linkage analysis in myotonic dystrophy. Cytogenet Cell Genet 1987; 46: 587 (Abst)

38 Myklebost O, Rogne S. A physical map of the apolipoprotein gene cluster on chromosome 19. Hum Genet 1988; 78: 244–247

39 Laberge C, Gaudet D, Morissette J, Moorjani S, Thibault MC. Linkage of myotonic dystrophy and APOE in a French-Canadian isolate. Cytogenet Cell Genet 1985; 40: 675

40 McKeithan TW, Rowley JD, Shows TB, Diaz M. Cloning of the chromosome translocation breakpoint junction of the t(14,19) in chronic lymphocytic leukaemia. Proc Natl Acad Sci 1987; 84: 9257–9260

41 Korneluk RG, MacLeod HL, McKeithan TW, Brook JD, McKenzie AE. A chromosome 19 clone from a translocation breakpoint shows tight linkage and linkage disequilibrium with myotonic dystrophy. Genomics 1989; 4: 146–151

42 Shaw DJ, Harley HG, Brook JD, McKeithan TW. Long-range restriction map of a region of human chromosome 19 containing the apolipoprotein genes, a CLL-associated translocation breakpoint and two polymorphic MluI sites. Hum Genet (submitted)

43 Nakamura Y, Lathrop M, O'Connell P, Leppert M, Lalouel JM, White R. A primary map of ten DNA markers and two serological markers for human chromosome 19. Genomics 1988; 3: 67–71

44 Coussens L, Parker PJ, Rhee L et al. Multiple, distinct forms of bovine and human protein kinase C suggest diversity in cellular signalling pathways. Science 1986; 233: 859–866

45 Harper PS, Bias WB, Hutchinson JR, McKusick VA. ABH secretor status of the fetus: A genetic marker identifiable by amniocentesis. J Med Genet 1971; 8: 438–440

46 Schrott HG, Omenn GS. Myotonic dystrophy: Opportunities for prenatal prediction. Neurology 1975; 25: 789–791

47 Estivill X, Scambler PJ, Wainwright BJ et al. Patterns of polymorphism and linkage disequilbrium for cystic fibrosis. Genomics 1987; 1: 257–263

48 Lindsay S, Bird AP. Use of restriction enzymes to detect potential gene sequences in mammalian DNA. Nature 1987; 327: 336–338

49 Monaco AP, Neve RL, Coletti-Feener C, Bertelson CJ, Kurnit DM, Kunkel LM. Isolation of candidate cDNAs for portions of the Duchenne muscular dystrophy gene. Nature 1986; 323: 646–650

50 Myers RM, Lumelsky N, Lerman LS, Maniatis T. Detection of single base substitutions in total genomic DNA. Nature 1985; 313: 495–498

British Medical Bulletin (1989) Vol. 45, No. 3, pp. 760–771
© The British Council 1989

Mitochondrial myopathies

A E Harding
I J Holt
Institute of Neurology, Queen Square, London, UK

The mitochondrial myopathies give rise to a diverse group of clinical syndromes, variably involving skeletal muscle and the central nervous system, with onset in childhood or adult life. In vitro studies of mitochondrial metabolism have identified a variety of functional defects of the respiratory chain, predominantly affecting complex I or complex III in adults, and complex IV in children. The increased incidence of maternal, as opposed to paternal, transmission in familial mitochondrial myopathy has led to the suggestion that these disorders may be caused by mutations of mitochondrial (mt) DNA. This hypothesis is derived from observations that mtDNA encodes subunits of the respiratory chain proteins and is exclusively maternally transmitted.

Analysis of muscle mtDNA shows two populations, one normal and the other deleted by up to nearly half its length, in about 40% of cases of mitochondrial myopathy. Only a single normal length population of mtDNA is seen in blood from these patients, and in blood and muscle from control subjects. Patients with muscle mtDNA deletions reported to date have all presented with progressive external ophthalmoplegia, including some with the Kearns-Sayre syndrome. They rarely have affected relatives. Deletions are not detected in cases of proximal myopathy alone, or those with adult onset syndromes predominantly affecting the central nervous system. There is no clear correlation between the deleted coding regions and the biochemical defects; even patients with seemingly identical muscle mtDNA deletions may be clinically and biochemically heterogeneous.

0007–1420/89/0045–0760/$10.00

CLINICAL FEATURES

The term mitochondrial myopathy (MM) is applied to a clinically and biochemically heterogeneous group of diseases which share the common feature of major structural abnormalities in skeletal muscle mitochondria. Ragged red fibres, containing peripheral and intermyofibrillar accumulations of abnormal mitochondria, seen with the modified Gomori trichrome stain,[1] are the major morphological hallmark of these disorders. These were initially observed commonly in patients presenting with syndromes of chronic progressive external ophthalmoplegia (CPEO) and/or proximal myopathy, often with weakness induced or enhanced by exertion. More recently, morphological evidence of mitochondrial dysfunction in muscle has been described in children and adults with complex multisystem disorders predominantly or exclusively affecting the central nervous system (CNS). These present with psychomotor retardation, dementia, pigmentary retinopathy, ataxia, seizures, movement disorders, stroke-like episodes, deafness, and peripheral neuropathy in various combinations. Involvement of other systems, such as the heart, endocrine glands, and haemopoietic tissues, has also been reported. The wide ranging clinical manifestations of the mitochondrial myopathies have recently been reviewed.[2,3]

It has been suggested that cases of MM can be classified into distinct syndromes on clinical grounds. These include: the Kearns-Sayre syndrome, a combination of CPEO and pigmentary retinopathy developing before the age of 20 years, associated with ataxia, cardiac conduction defects, and increased cerebrospinal fluid protein concentrations;[4] the syndrome of myoclonus epilepsy with ragged red fibres (MERRF);[5] and a further syndrome of mitochondrial encephalopathy, lactic acidosis and stroke-like episodes (MELAS).[6] Petty and colleagues,[3] reporting a series of 66 patients with MM, felt that these syndromes were not specific even on clinical grounds, as there is considerable overlap between them, but that they merely represent combinations of some of the more striking features of the mitochondrial myopathies.

BIOCHEMICAL FEATURES

Many patients with MM have a pathological increase in serum lactate concentration during and after exercise, suggesting a defect of aerobic metabolism in muscle mitochondria, and defects of the mitochondrial respiratory chain have been demonstrated in the majority of cases investigated.

The main products of oxidation of pyruvate and fatty acids in the mitochondrial matrix are carbon dioxide and NADH. NADH is the principal substrate for the respiratory chain enzymes which are situated in the inner mitochondrial membrane. Electrons from NADH (and $FADH_2$) pass along the electron transfer chain, gradually releasing energy which pumps protons across the inner membrane. This process utilizes oxygen and creates an electrochemical proton gradient across the inner membrane which is largely used to drive the conversion of ADP and phosphate to ATP by the enzyme mitochondrial ATPase.[7]

The respiratory chain comprises four major enzyme complexes which are embedded in the inner mitochondrial membrane (Fig.1). Most of the electron carriers in the chain contain metal atoms (haem or non-haem iron, and copper) which are tightly bound to a protein surface. The NADH-Coenzyme Q (CoQ) reductase complex (complex I) consists of about 26 polypeptides, including eight iron sulphur (FeS) proteins, and accepts electrons

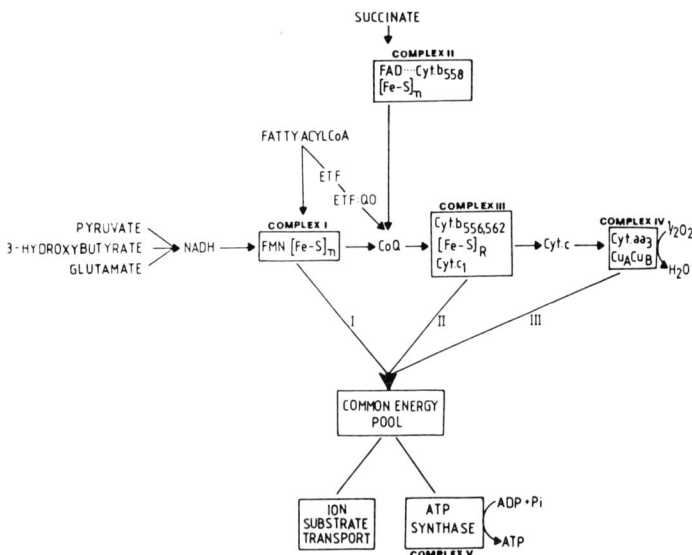

Fig. 1 Diagram representing the mitochondrial respiratory chain and oxidative phosphorylation system (*reproduced from*: Rosenberg R N, Harding A E, eds. The Molecular Biology of Neurological Disease. London: Butterworths, 1988, by courtesy of the publishers).

from NADH before transferring them to complex III (ubiquinol-cytochrome c reductase) via a small lipid soluble molecule, ubiquinone (CoQ). $FADH_2$ feeds into the chain at complex III via complex II (succinate-CoQ reductase). The 11 or so subunits of complex III include cytochromes b and c_1, and two FeS proteins. Electrons pass from ubiquinone to cytochrome c via complex III, and then to the cytochrome oxidase complex (complex IV) consisting of 13 subunits including cytochromes a and a_3. Cytochrome oxidase forms two water molecules from four electrons transferred from cytochrome c and oxygen.[7]

The electrochemical proton gradient created by electron transfer along the respiratory chain drives the production of ATP from ADP and phosphate by the enzyme complex ATP synthetase (complex V) bound to the inner mitochondrial membrane. The energy supplied by the gradient is also used to transport pyruvate and other mitochondrial enzyme substrates, and nuclear encoded mitochondrial proteins, into the mitochondrial matrix.[8]

Human muscle mitochondria can be isolated from large biopsy specimens in sufficient amounts for polarographic studies of oxidation and phosphorylation, and for the determination of cytochrome spectra.[9] These studies have identified a variety of defects of the respiratory chain and oxidative phosphorylation system in patients with MM, all of which are associated with a wide range of clinical syndromes. Conversely, symptoms and signs may be similar in patients with different biochemical defects.[2,3,10]

Adult patients with MM most frequently have defects of the respiratory chain localized to complex I (NADH-CoQ reductase).[3,10] Approximately half of the reported cases have had myopathy alone, and half a multisystem disorder involving muscle, the retina, and/or the CNS.[10] Complex I defects may also be seen in cases of infantile lactic acidosis with myopathy.[11] Complex II deficiency is extremely rare. The clinical features observed in association with complex III defects in adults are also variable, usually consisting of proximal myopathy with CPEO, although CNS disease has been described.[10]

Several clinical syndromes occur in association with cytochrome c oxidase (complex IV) deficiency, including: (1) fatal infantile mitochondrial myopathy with lactic acidosis, often combined with the de Toni-Fanconi-Debre renal syndrome and sometimes with cardiac or hepatic involvement; (2) benign infantile mitochondrial myopathy secondary to reversible cytochrome c oxidase deficiency; and (3) subacute necrotizing encephalomyelopathy (Leigh's syndrome).[2]

A number of cases of MM and defects involving more than one complex have been described, as has one example of the Kearns-Sayre syndrome with mitochondrial ATP synthetase deficiency.[2,3]

GENETIC ASPECTS

Families containing more than one individual with MM have been described, but the majority of cases are not familial. In a series of 71 patients, 18% of index cases had similarly affected relatives.[12] No consistent pattern of inheritance was evident for any of the clinical syndromes or identified defects of mitochondrial metabolism, in either this study or other reports. Some pedigrees suggest autosomal recessive or dominant inheritance; no convincing X-linked pedigrees have been described. It is clear that, when individuals are affected in more than one generation, maternal transmission to offspring is far more frequent than paternal transmission (in a ratio of approximately 9:1).[12] Hudgson and colleagues[13] and Egger and Wilson[14] suggested that this could be explained on the basis of mitochondrial inheritance, as mitochondrial DNA is exclusively maternally transmitted. Support for this hypothesis comes from the fact that the majority of patients with MM have biochemical defects localized to complex I, III, or IV of the respiratory chain, all of which contain subunits encoded by mitochondrial DNA.

If MM is mitochondrially inherited in pedigrees indicating maternal transmission, all the offspring of affected females should be affected, and only about half of them are.[12] One explanation for this is that some individuals carrying the abnormal mitochondrial genotype do not express it clinically or histologically. The diagnosis of MM is sometimes difficult to confirm or exclude with certainty; clinically affected individuals with normal skeletal muscle biopsies and in vitro studies of mitochondrial metabolism have been described in some families.[3,15] An alternative explanation for reduced penetrance of an abnormal mitochondrial genotype in any single maternal line is that ova contain a heterogeneous population of mtDNA molecules.

THE MITOCHONDRIAL GENOME

Mammalian mitochondria each contain 5–10 circular DNA molecules which are double stranded and about 16.5 kilobases (kb) in length.[8,16,17] The human mitochondrial genome has been sequenced;[18] it differs from nuclear DNA to some extent in its genetic code which, for example, dictates that UGA reads tryptophan instead of a stop codon, and also because it contains very little non-coding sequence. Each strand of mitochondrial DNA (mtDNA) appears to be transcribed from a single promotor site, and then processed. The heavy (H) strand transcripts consist of two ribosomal RNAs, 14 tRNAs, and 12 protein coding sequences, and the light (L) strand codes for 8 tRNAs and one protein coding sequence. The mitochondrial protein coding transcripts are not capped at their 5′ end but their 3′ ends are polyadenylated by mitochondrial poly-A polymerase.[16,18-20] Mitochondria divide at a rate appropriate to that of division of their parent cells, and in most instances mtDNA molecules replicate with every cell cycle. mtDNA contributes about 1% of total cellular DNA.

mtDNA encodes for 13 of the 67 or so subunits of the mitochondrial respiratory chain and oxidative phosphorylation system (Fig.1): seven subunits of complex I; cytochrome b (complex III); subunits I, II, and III of cytochrome oxidase (complex IV); and subunits 6 and 8 of ATP synthetase.[16,21,22] The nuclear genome encodes the remaining polypeptides in the respiratory chain, and also controls their transport into the mitochondrion by synthesizing 'leader' peptides which appear to direct the proteins to sites of adhesion between the inner and outer mitochondrial membrances prior to transport into the matrix.[8,23] Replication, transcription, and translation of the mitochondrial genome are also dependent on nuclear products such as ribosomal proteins, the enzymes involved in mtDNA replication, and RNA polymerase.[19]

mtDNA has been shown to be exclusively maternally transmitted in many species, including man.[24,25] A few paternal mitochondria may penetrate the ovum, but these appear to degenerate subsequently.[26] Maternal inheritance of mtDNA has been confirmed in recent years using molecular genetic techniques. mtDNA from all individuals in a single maternal line shows the same pattern of fragments after digestion with restriction endonucleases,[24,27] although extensive mtDNA nucleotide sequence divergence occurs between different maternal lines.[27,28] The mutation rate of mtDNA is high, and restriction mapping of mtDNA in different human populations can be used to trace their origins.[28]

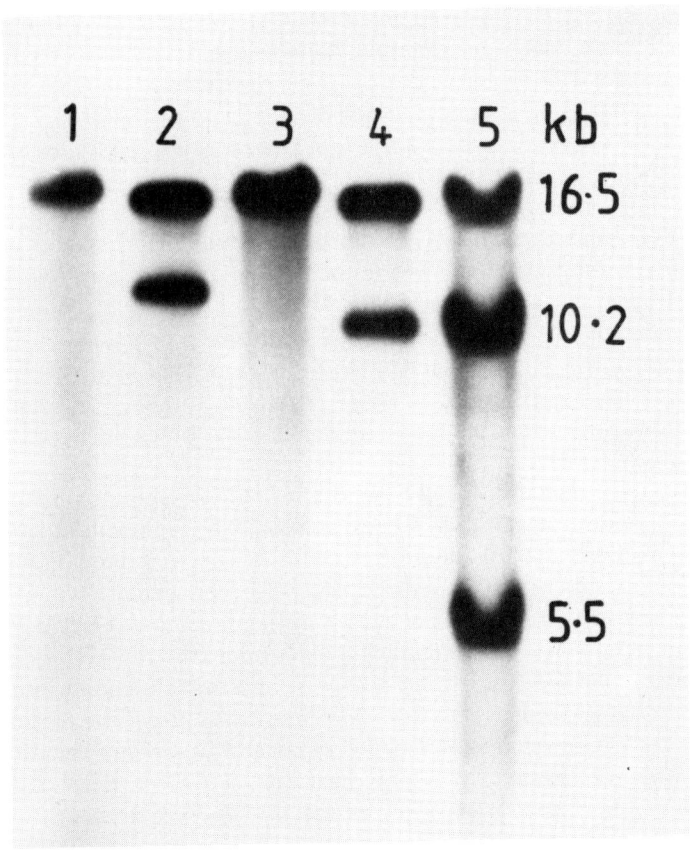

Fig. 2 Hydridization patterns of ^{32}P-labelled HeLa cell mtDNA to mtDNA digested with BamHI from: blood from a patient with mitochondrial myopathy (lane 1); muscle from the same patient (lane 2); and blood (lane 3) and muscle (lane 4) from a second patient. Size marker fragments of control mtDNA are shown in lane 5. BamHI cleaves the mitochondrial genome at one site, thus making it linear. In blood only one fragment (16.5 kb) is seen, but in muscle (lanes 2, 4) there are two. The lower bands represent mtDNAs bearing deletions of about 5–6 kb.

Despite the high mutation rate of mtDNA, it is generally thought that all the mtDNA in an individual normal human is identical, although extensive sequencing studies of mtDNA from different tissues have not been described. mtDNA heteroplasmy has been demonstrated in *Drosophila*[29] and a single maternal line of Holstein cows.[30]

STUDIES OF THE MITOCHONDRIAL GENOME IN MM

Restriction endonuclease analysis of mtDNA from blood in families with MM has excluded major deletions of leukocyte mtDNA in patients, or any differences in restriction fragment patterns between normal and abnormal individuals in the same maternal line.[31-33] This approach does not exclude the presence of small deletions or pathologically significant mutations outside restriction sites, and there is only about a 10% chance of detecting the latter using about 30 restriction endonucleases.

Holt and colleagues[34] showed that 9 of 25 patients with MM had two populations of mtDNA in muscle, one of which was deleted by up to 7 kb (Fig.2). None of these 9 cases had detectably abnormal leukocyte mtDNA. The proportion of abnormal mtDNAs in muscle ranged from 20–70%. Further cases of MM with deletions of a proportion of muscle mtDNAs have been reported subsequently.[35,36]

To date we have screened muscle mtDNA from 71 patients and 28 have a deleted population of mtDNA in muscle (Holt et al., unpublished observations). The deleted region has been mapped in 16; 11 have similar if not identical deletions extending over 5 kb within the region 8000–13600 bp (Fig.3). All of our 28 patients

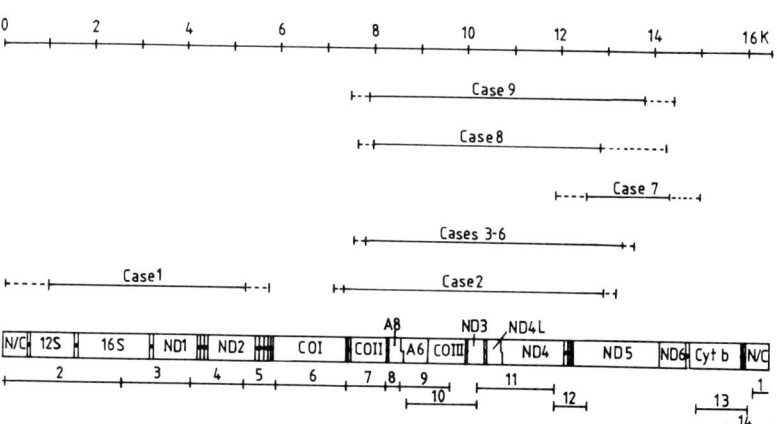

Fig. 3 Linearized map of the mitochondrial genome showing location of coding regions, and the extent of the deleted regions in 9 cases of mitochondrial myopathy. The interrupted lines at each end of the deletions represent their upper and lower limits, defined by restriction mapping and hybridization patterns to a series of 14 small mtDNA fragments which map as shown at the bottom of the figure. ND = NADH CoQ reductase (complex I), A = ATPase, CO = cytochrome oxidase, 12S and 16S = ribosomal RNAs; the symbol o indicates a tRNA.

with deletions of muscle mtDNA presented clinically with CPEO and proximal myopathy, compared to 12 of 43 patients without detectable deletions, and 8 of these had the Kearns Sayre syndrome of CPEO, short stature, retinopathy, ataxia and cardiac conduction defects (Table 1).[4] None of the 11 patients presenting with proximal myopathy without CPEO or the 19 with predominant CNS disease had deletions of muscle mtDNA, and a single population of normal length mtDNA was detected in the brain of one of the latter cases.[37]

Apart from the patients with Kearns-Sayre syndrome, the cases with demonstrable deletions were relatively mildly affected, and this was reflected by biochemical studies of mitochondria which generally showed a mild reduction in respiratory capacity. The site of the respiratory chain defect was variable, even in patients with seemingly identical deletions, and sometimes poorly localized; there was no obvious correlation between the site of the biochemical defect and the deleted mitochondrially encoded subunits.

These observations demonstrate conclusively that mtDNA heteroplasmy can occur in man, and that human disease may be associated with defects of the mitochondrial genome. The highly variable clinical manifestations of MM, and the incomplete penetrance in the offspring of female patients, are adequately explained by mtDNA heteroplasmy between different tissues. The survival of deleted mtDNA molecules in muscle is compatible with the observation that the number of muscle fibres does not increase significantly after early fetal life.[38] The same may apply to CNS tissue in some cases of MM presenting with predominant CNS disease. On the other hand, frequent cell division in leukocyte precursors should select against cells containing genetically defective mitochondria.

Table 1   Clinical and genetic features of patients with mitochondrial myopathy with and without deletions of muscle mitochondrial DNA

	No. of cases	
	mtDNA deletion (28)	no mtDNA deletion (43)
Affected relatives	2	10
CPEO/limb weakness	20	11
Kearns Sayre syndrome	8	2
Limb weakness, no CPEO	0	11
Predominantly CNS disease	0	19

The origin of muscle mtDNA deletions is unclear, in terms of both timing and molecular mechanism. Only two of our patients with muscle mtDNA deletions had affected relatives; it seems probable that the deletions in most cases arose during oogenesis and that random partitioning of the two populations of mtDNA occurred during fetal development. There are no clues as to the molecular basis of mtDNA deletions. As 11 of our cases have very similar, if not identical, deletions, it could be predicted that these arose as a result of a recombination hotspot.

The absence of detectable muscle mtDNA deletions in about two thirds of patients with MM does not exclude the possibility of mtDNA mutations as the basis for their disease. The limitations of restriction endonuclease analysis mean that deletions of less than approximately 100 bp, or point mutations, have not been excluded. Nevertheless, on statistical grounds alone, it is unlikely that all cases of MM are due to defective mitochondrial genes, as the nuclear genome codes for the majority of the respiratory chain subunits, as well as controlling their transport into the mitochondrion and subsequent assembly into functional enzyme complexes; replication, transcription, and translation of the mitochondrial genome are also dependent on the nucleus.[19] Evidence that MM may be caused by mutant nuclear genes is provided by Schapira and colleagues,[39] who showed that some patients with complex I defects have a specific deficiency of the 24 kD FeS protein which is a nuclear product.

ACKNOWLEDGEMENTS

Much published and unpublished work reviewed in this article was done or made possible by John Morgan-Hughes, John Clark, Mark Cooper, and Anthony Schapira. We thank Marjorie Gilbert for technical assistance, and the Muscular Dystrophy Group of Great Britain and Northern Ireland, the Research Trust for Metabolic Diseases of Childhood, and the Brain Research Trust for financial support.

REFERENCES

1 Olson W, Engel WK, Walsh GO, Einaugler R. Oculocraniosomatic neuromuscular disease with 'ragged red' fibres. Arch Neurol 1972; 26: 193–211
2 DiMauro S, Bonilla E, Zeviani M, Nakagawa M, DeVivo DC. Mitochondrial myopathies. Ann Neurol 1985; 17: 521–538
3 Petty RKH, Harding AE, Morgan-Hughes JA. The clinical features of mitochondrial myopathy. Brain 1984; 109: 915–938
4 Berenberg RA, Pellock JM, DiMauro S et al. Lumping or splitting? 'Ophthalmoplegia-plus' or Kearns-Sayre syndrome? Ann Neurol 1977; 1: 37–54

5 Fukuhara N, Tokiguchi S, Shirakawa S, Tsubaki T. Myoclonus epilepsy associated with ragged red fibres (mitochondrial abnormalities): disease entity or syndrome. J Neurol Sci 1980; 47: 117–133

6 Pavlakis SG, Phillips PC, DiMauro S, DeVivo DC, Rowland LP. Mitochondrial myopathy, encephalopathy, lactic acidosis, and strokelike episodes: a distinctive clinical syndrome. Ann Neurol 1984; 16, 481–488

7 Hatefi Y. The mitochondrial electron transport and oxidative phosphorylation system. Ann Rev Biochem 1985; 54: 1015–1069

8 Alberts B, Bray D, Lewis J, Raff M, Roberts K, Watson JD. Molecular Biology of the Cell. New York: Garland, 1983

9 Morgan-Hughes JA, Darveniza P, Kahn SN et al. A mitochondrial myopathy characterised by a deficiency in reducible cytochrome b. Brain 1977; 100: 617–640

10 Morgan-Hughes JA, Hayes DJ, Cooper M, Clark JB. Mitochondrial myopathies: deficiencies localised to complex I and complex III of the respiratory chain. Biochem Soc Trans 1985; 13: 648–650

11 Robinson BH, Ward J, Goodyer P, Baudet A. Respiratory chain defects in the mitochondria of cultured skin fibroblasts from three patients with lacticacidaemia. J Clin Invest 1986; 77: 1422–27

12 Harding AE, Petty RKH, Morgan-Hughes JA. Mitochondrial myopathy: a genetic study of 71 cases. J Med Genet 1988; 25: 528–535

13 Hudgson P, Bradley WG, Jenkison M. Familial mitochondrial myopathy. A myopathy with disordered oxidative metabolism in muscle fibres. Part 1. Clinical, electrophysiological and pathological findings. J Neurol Sci 1972; 16: 343–370

14 Egger J, Wilson J. Mitochondrial inheritance in a mitochondrially mediated disease. N Engl J Med 1983; 309: 142–145

15 Rosing HS, Hopkins LC, Wallace DC et al. Maternally inherited mitochondrial myopathy and myoclonic epilepsy. Ann Neurol 1985; 17: 228–237

16 Attardi G. The elucidation of the human mitochondrial genome: a historical perspective. BioEssays 1986; 5: 34–39

17 Tzagoloff A, Myers AM. Genetics of mitochondrial biogenesis. Ann Rev Biochem 1986; 55: 249–285

18 Anderson S, Bankier AT, Barrell BG et al. Sequence and organisation of the human mitochondrial genome. Nature 1981; 290: 457–465

19 Attardi G. Organization and expression of the mammalian mitochondrial genome: a lesson in economy. Trends Biochem Sci 1981; 6: 86–89, 100–103

20 Clayton DA. Transcription of the mammalian mitochondrial genome. Ann Rev Biochem 1984; 53: 573–594

21 Chomyn A, Mariottini P, Cleeter MWJ et al. Six unidentified reading frames of human mitochondrial DNA encode components of the respiratory chain NADH dehydrogenase. Nature 1985; 314: 592–597

22 Chomyn A, Mariottini P, Cleeter MWJ et al. Functional assignment of the unidentified reading frames of human mitochondrial DNA. In: Quagliariello E et al., eds. Achievements and Perspectives of Mitochondrial Research, volume II: Biogenesis, Amsterdam: Elsevier, 1985: pp. 259–275

23 Neupert W, Schatz G. How proteins are transported into mitochondria. Trends Biochem Sci 1981; 6: 1–4

24 Giles RE, Blanc H, Cann HM, Wallace DC. Maternal inheritance of human mitochondrial DNA. Proc Natl Acad Sci USA 1980; 77: 6715–6719

25 Lansman RA, Avise JC, Huettel MD. Critical experimental test of the possibility of 'paternal leakage' of mitochondrial DNA. Proc Natl Acad Sci 1983; 80: 1969–1971

26 Fine PEM. Mitochondrial inheritance and disease. Lancet 1978, ii: 659–662

27 Brown WM. Polymorphism in mitochondrial DNA of humans as revealed by restriction endonuclease analysis. Proc Natl Acad Sci USA 1980; 77: 3605–3609

28 Cann RL, Stoneking M, Wilson AC. Mitochondrial DNA and human evolution. Nature 1987; 325: 31–36
29 Solignac M, Monnerot M, Mounolou J-C. Mitochondrial DNA heteroplasmy in Drosophila mauritiana. Proc Natl Acad Sci USA 1983; 80: 6942–6946
30 Hauswirth WW, Laipis PJ. Mitochondrial DNA polymorphism in a maternal lineage of Holstein cows. Proc Natl Acad Sci USA 1980; 79: 4686–4690
31 Wallace DC, Singh G, Hopkins LC, Novotny EJ. Maternally inherited diseases of man. In: Quagliariello E et al., eds. Achievements and Perspectives of Mitochondrial Research, volume II: Biogenesis. Amsterdam: Elsevier, 1985: p. 427–436
32 Holt IJ, Harding AE, Morgan-Hughes JA. Mitochondrial DNA polymorphism in mitochondrial myopathy. Hum Genet 1988; 79: 53–57
33 Poulton J, Turnbull DM, Mehta AB, Wilson J, Gardiner RM. Restriction enzyme analysis of the mitochondrial genome in mitochondrial myopathy. J Med Genet 1988; 25: 600–605
34 Holt IJ, Harding AE, Morgan-Hughes JA. Deletions of mitochondrial DNA in patients with mitochondrial myopathies. Nature 1988; 331: 717–719
35 Lestienne P, Ponsot G. Kearns-Sayre syndrome with muscle mitochondrial DNA deletion. Lancet 1988; i: 885
36 Zeviani M, Moraes CT, DiMauro S et al. Deletions of mitochondrial DNA in Kearns-Sayre syndrome. Neurology 1988; 38: 1339–1346
37 Holt IJ, Cooper JM, Morgan-Hughes JA, Harding AE. Deletions of muscle mitochondrial DNA. Lancet 1988; i: 1462
38 Stickland NC. Muscle development in the human fetus as exemplified by m. sartorius: a quantitative study. J Anat 1981; 132: 557–579
39 Schapira AHV, Cooper JM, Morgan-Hughes JA et al. Molecular basis of mitochondrial myopathies: polypeptide analysis in complex I deficiency. Lancet 1988; i: 500–503

British Medical Bulletin (1989) Vol. 45, No. 3, pp. 772–787
© The British Council 1989

Emery-Dreifuss muscular dystrophy and other related disorders

A E H Emery
Medical School, University of Edinburgh, Edinburgh, UK

There are some 30 or so different forms of muscular dystrophy which are conveniently classified according to the mode of inheritance. Emery-Dreifuss X-linked muscular dystrophy is characterized by the triad of: (1) *early* contractures of the elbows, Achilles tendons and post-cervical muscles; (2) slowly progressive muscle wasting and weakness with a humero-peroneal distribution in the early stages; and (3) a cardiomyopathy usually presenting as heart-block. The insertion of a cardiac pace-maker can be life saving and therefore the recognition of the condition is essential. The responsible gene has been localized to Xq28.

The autosomal recessive dystrophies are classified into congenital forms (the Fukuyama type is particularly common in Japan); a childhood form (similar to Duchenne) which occurs frequently in certain inbred communities; and adult onset limb girdle dystrophy.

The autosomal dominant dystrophies are classified on the distribution of predominant muscle weakness into facioscapulohumeral, scapuloperoneal (with or without early contractures and cardiomyopathy), proximal, distal and ocular forms.

The basic biochemical defects and the localizations of the responsible genes are as yet unknown in any of the autosomal recessive or autosomal dominant dystrophies.

The muscular dystrophies may be defined as a group of inherited disorders characterized by progressive muscle wasting and weak-

0007–1420/89/0045–0772/$10.00

ness but with no evidence of involvement of the peripheral nerves or spinal cord and no myotonia. The muscle histology is distinctive with variation in muscle fibre size, fibre necrosis and phagocytosis and later replacement by fat and connective tissue. Some 20 different types of muscular dystrophy have now been identified (Table 1) which vary considerably in their clinical presentation and severity. At least in Britain and the United States the X-linked forms are commonest. Since the Duchenne and Becker forms of muscular dystrophy are dealt with elsewhere, here the Emery-Dreifuss form (EDMD) will be discussed in detail and relevant features of the other dystrophies also mentioned.

Table 1   Clinical and genetic classification of the muscular dystrophies

1.	*X-linked dystrophies*		
	a.	Proximal	
		i.	Duchenne
		ii.	Becker
		iii.	Mabry
	b.	With early contractures and cardiomyopathy (Emery-Dreifuss)	
	c.	Scapuloperoneal with mental retardation (Bergia)	
	d.	Quadriceps	
		Quadriceps myopathy ?	
2.	*Autosomal recessive dystrophies*		
	a.	Congenital forms	
		i.	rapidly progressive
		ii.	slowly progressive (numerous variants)
	b.	Childhood form	
	c.	Adult form	
		Limb girdle dystrophy	
	d.	Scapulohumeral	
	e.	Quadriceps myopathy	
3.	*Autosomal dominant dystrophies*		
	a.	Facioscapulohumeral	
	b.	With early contractures and cardiomyopathy	
	c.	Scapuloperoneal	
	d.	Proximal	
		i.	dominant limb girdle dystrophy
		ii.	hereditary myopathy limited to females
		iii.	hereditary myopathy limited to males
	e.	Distal	
		i.	childhood form
		ii.	adult form
	f.	Ocular	
		i.	ocular form
		ii.	oculopharyngeal forms (AD,AR)

EMERY-DREIFUSS MUSCULAR DYSTROPHY (EDMD)

The distinctive clinical features of this relatively benign form of X-linked dystrophy were delineated in 1966,[1] but the disorder may well have been recognized earlier in two affected brothers.[2] There have now been a further 20 reports of the disorder in 22 families with a total of 177 affected individuals.[3,4] The disorder is characterized by the triad of: (1) *early* contractures of the elbows, Achilles tendons and post-cervical muscles; (2) slowly progressive muscle wasting and weakness with a humero-peroneal distribution in the early stages; and (3) a cardiomyopathy usually presenting as heart block.

Early contractures

Muscle contractures are a frequent finding in the later stages of any form of dystrophy and spinal muscular atrophy. However the development of contractures *before* the development of any significant muscle weakness seems to be limited only to this form of dystrophy. Involvement of the elbows results in the arms being carried in a semiflexed position often from childhood, and the tightening of the Achilles tendons results in the affected individual walking on his toes. The contractures of the post-cervical muscles result in limitation of neck flexion which may pass unnoticed unless looked for specifically. Later forward flexion of the thoracic and lumbar spine also becomes limited.

Muscle wasting and weakness

Early in the course of the disease muscle weakness predominantly affects the biceps and triceps muscles and at this stage the anterior tibial and peroneal muscles are also affected. Later weakness of the hip and knee extensors and the proximal upper limb musculature also develops. In the lower limbs the distal muscles are affected before the proximal muscles. Thus the distribution of muscle weakness may be described as humero-peroneal at the beginning, and as scapulo-humero-pelvo-peroneal later in the course of the disease.

Cardiomyopathy

This usually presents as heart block and may develop in the early 20's or even in adolescence.[5] Before symptoms develop it can be recognized by sinus bradycardia and electrocardiographic evidence of prolongation of the PR interval. Syncope attacks are not uncommon and later frank heart block develops and is a frequent cause of death. Provided the diagnosis is made sufficiently early the insertion of a cardiac pacemaker can be life saving.[6-11] Early recognition of the condition is therefore essential (Fig. 1).

Biochemistry

The serum level of creatine kinase (SCK) is usually moderately raised but even in the early stages of the disease never approaches the grossly elevated levels found in the Duchenne and Becker

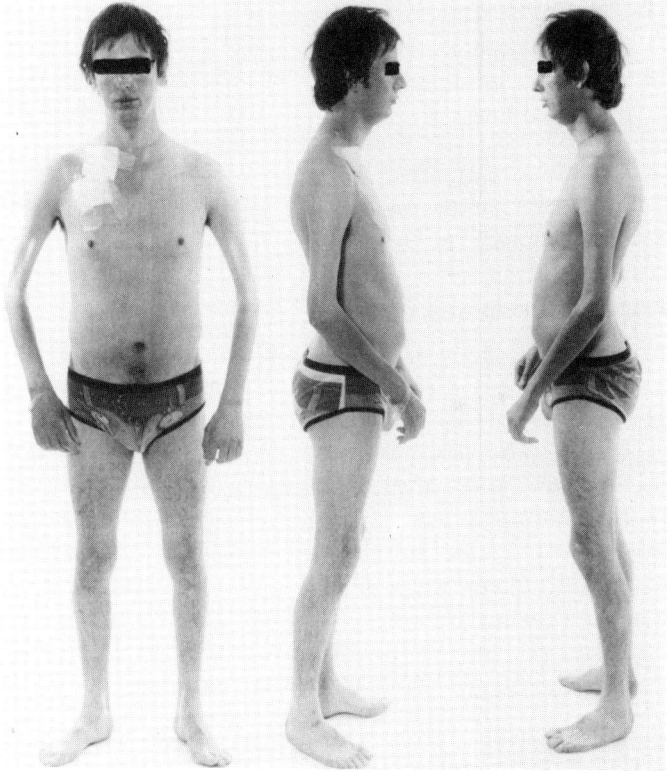

Fig. 1 A 17 year-old boy with Emery-Dreifuss muscular dystrophy. Note the flexion contractures of the elbows and wasting of the lower legs. A cardiac pacemaker has been inserted. (*From* Emery A E H 1988[4] 'Duchenne Muscular Dystrophy', Oxford University Press, with permission.)

forms of dystrophy. Autopsy studies have shown that the spinal cord is normal and that the ventral spinal roots are intact.[12]

Prognosis

There does not appear to be any significant intellectual deficit in this disorder which is relatively slowly progressive. Unless a cardiomyopathy intervenes most affected individuals may expect to survive into middle age with varying degrees of incapacity.

Molecular genetics

The responsible gene is linked to colour blindness[13] and to DNA markers located around Xq28.[14-17] Since the loci for Duchenne and Becker muscular dystrophies are located at Xp21, the gene for EDMD cannot be allelic with these disorders. Presumably therefore EDMD has a different biochemical basis. In fact the protein 'dystrophin', found to be absent in muscle in Duchenne muscular dystrophy and significantly reduced in Becker muscular dystrophy, has been found to be normal in EDMD.[18,19] That the responsible gene has now been localized may soon lead to its isolation, and ultimately to characterization of its product.

Carrier detection and prenatal diagnosis

Female carriers rarely if ever exhibit any significant muscle weakness, but bradycardia, arrhythmias and significantly prolonged PR intervals are frequent findings, and frank heart block may occur in otherwise healthy carriers.[3,20-22] A proportion of known carriers have slightly raised SCK levels and this may be helpful in identifying such women in affected families. However closely linked DNA markers can also be used for carrier detection as well as for prenatal diagnosis.

AUTOSOMAL DYSTROPHY WITH EARLY CONTRACTURES AND CARDIOMYOPATHY

There have been several reports of an autosomal dystrophy with essentially scapulo-peroneal weakness associated with early contractures of the elbows, Achilles tendons and post-cervical muscles and a cardiomyopathy.[3] With one possible exception all have been inherited as an autosomal dominant trait. Clinically the disorder is very similar to the X-linked form but may differ in that the former is perhaps more variable in age at onset and degree of severity.[23] Another autosomal dominant disorder described in the literature[24] appears to be similar but distinct because cardiac conduction defects are *not* a feature.

OTHER X-LINKED DYSTROPHIES

Two other forms of relatively benign X-linked dystrophy have been described but so far only in single families. The first report[25] described a large family from Kentucky with an X-linked dystrophy with prolonged survival and associated with relatively severe disability and myocardial involvement (Mabry type). There appear to have been no reports of any other families with exactly the same disorder. More recently a family in Texas has been reported with another relatively benign (apparently) X-linked dystrophy.[26] The muscle weakness was scapulo-peroneal in distribution and was associated with a lethal cardiomyopathy (Bergia type). In these regards therefore this disorder somewhat resembles EDMD but differs in that early muscle contractures are *not* a feature and all three affected individuals were markedly mentally retarded. Some distinguishing clinical features between the various forms of X-linked dystrophy are summarized in Table 2.

A syndrome characterized by scapulo-peroneal weakness but without facial weakness, can be inherited as an autosomal dominant trait.[27,28] But onset is usually in adult life, *early* contractures do *not* occur and cardiac conduction defects are *not* a consistent feature. Furthermore this pattern of muscle involvement, inherited either as an autosomal dominant or autosomal recessive trait, is usually neurogenic in origin.[29]

Finally a form of dystrophy predominantly affecting the quadriceps muscle, and therefore sometimes referred to as 'quadriceps myopathy' may be inherited as an X-linked recessive trait in some families[30] or possibly as an autosomal recessive trait in others. However the relationship of this disorder to the 'limb girdle syndrome' and its nosological status in general is not clear.

AUTOSOMAL RECESSIVE DYSTROPHIES

For the sake of convenience these disorders can be subdivided mainly on the basis of age at onset: congenital forms where the disease is present at birth, childhood forms where onset is in early

Table 2 Some distinguishing clinical features between the various forms of X-linked muscular dystrophy

Type	Onset (usual)	Course	Predominant weakness	Contractures (early)	Mental handicap	Cardiomyopathy
Duchenne	<5	severe	proximal	−	+/−	+
Becker	5–25	benign	proximal	−	−?	−
Emery-Dreifuss	early childhood	benign	humero-peroneal	+	−	+
Mabry	adolescence	benign	proximal	−	−	+
Bergia	early childhood	benign ?	scapulo-peroneal	−	+	+

childhood, and an adult form (limb girdle muscular dystrophy) where onset is in adolescence or later. The basic biochemical defects in these disorders and their chromosomal locations are unknown at present.

Congenital muscular dystrophy

This is a very heterogeneous group of disorders for which at present there is no satisfactory classification. In some cases, not associated with any features other than muscular dystrophy, progression is rapid and death occurs in infancy or early childhood. However, in other families with congenital muscular dystrophy the course of the disease has been either non-progressive or only slowly progressive. But the distinction between severe and more benign forms has been questioned and it is often difficult to predict when the diagnosis is first made what the ultimate prognosis may be.[31] A particularly frequent muscular dystrophy in Japan is an autosomal recessive congenital form which is associated with mental handicap.[32,33] Other congenital forms have been described in association with infantile cataracts and hypogonadism,[34] CNS malformations,[35] congenital heart disease,[36] and with cardio-respiratory failure (so-called Eichsfelt type of congenital muscular dystrophy).[37] A non-progressive form of congenital muscular dystrophy has been described as part of a syndrome which includes high arched palate, hyperhidrosis, protrusion of the calcaneum and a tendency to recurrent upper respiratory tract infections. This disorder, sometimes referred to as Ullrich syndrome or congenital hypotonic-sclerotic muscular dystrophy, is also inherited as an autosomal recessive trait.[38,39]

Autosomal recessive muscular dystrophy of childhood

An autosomal recessive form of muscular dystrophy can affect both boys and girls.[40-42] Clinically the condition is very similar to Duchenne muscular dystrophy though the former may be somewhat milder, affected children often not becoming confined to a wheelchair until their early teens or even later. Electrocardiographic evidence of right ventricular preponderance (tall R waves in V_1) is found only in the X-linked Duchenne form.[43,44] In Britain and North America the X-linked form would seem to be at least 20 times commoner than the autosomal recessive form.[45] The latter, however, has been reported in Arab communities in Tunisia[46] and the Sudan.[47] An autosomal recessive form also occurs in certain inbred communities in North America[48-50] which originated in Switzerland where this form of muscular dystrophy seems to be particularly prevalent.[51]

The occurrence of an autosomal recessive form of dystrophy which clinically closely resembles the Duchenne form can result in errors in counselling and could be a confounding factor in studying the aetiology and possible pathogenesis of Duchenne muscular dystrophy. Any affected girl should be karyotyped (to exclude an X chromosome defect or X/autosome translocation) and all affected boys and girls should have an ECG examination for evidence of right ventricular preponderance.

Limb girdle muscular dystrophy

The term 'limb girdle muscular dystrophy' was used[52] for a disorder characterized by proximal muscle wasting and weakness affecting both males and females with onset usually in the second and third decades of life and subsequently slowly progressive. However, it is now clear that such a definition would also include many other disorders, at least when pelvifemoral weakness predominates (Table 3), and in fact it has been questioned[53] whether limb girdle muscular dystrophy as such really exists. Certainly if this diagnosis is to be considered then it is important to exclude various other possibilities on the basis of clinical, family and laboratory studies. When this is done the diagnosis of limb girdle muscular dystrophy may be entertained but then appears to be a rare disorder[54] with a prevalence of around 3×10^{-6}. Onset is in

Table 3 Possible causes of the 'limb girdle syndrome' of adulthood

1. *Muscular dystrophies*
   a. Benign X-linked forms
   b. Manifesting carrier of Duchenne form
   c. Dominant proximal and scapuloperoneal forms
2. *Spinal muscular atrophies*
      (e.g.) Kugelberg-Welander form
3. *Myopathies*
   a. Acquired
      (e.g.) sarcoidosis, thyrotoxicosis, metabolic bone disease, acromegaly
   b. Hereditary
      (e.g.) central core disease, acid maltase deficiency, carnitine deficiency, 'nemaline' and 'mitochondrial' myopathies
4. *Polymyositis*
5. *Drug induced*

early adult life and the pectoral girdle musculature is often affected first with the pelvic girdle musculature not becoming overtly affected until several years later. The distal musculature is not affected, calf pseudohypertrophy is rare, the heart is unaffected and intelligence is normal. The serum level of creatine kinase is normal or only slightly raised.

The disorder can be inherited as an autosomal recessive trait[51] but many cases are sporadic. It is in these latter cases[53] that it has been pointed out, '... (they) may well be, if male, isolated cases of Becker dystrophy, and if female, manifesting carriers of the Duchenne gene'.

Finally, a rare adult onset 'scapulo-humeral' type of dystrophy has been described. Here weakness first affects the proximal upper limb musculature and only later affects the lower limbs and face. It appears to be relatively common in Switzerland.[51]

AUTOSOMAL DOMINANT DYSTROPHIES

This group of disorders can be subdivided on the basis of the distribution of predominant muscle weakness. As in the case of the autosomal recessive dystrophies, the basic biochemical defects and their chromosomal locations are unknown at present.

Facioscapulohumeral muscular dystrophy

This is the commonest of the autosomal dominant dystrophies. Classically the disorder, which first becomes evident usually in adolescence, is characterized by weakness of the facial and pectoral girdle muscles. The anterior tibial (peroneal) muscles often become affected early on. It is usually only after several years that the pelvic girdle musculature may also become affected. However there is considerable variation in severity even within families. Some individuals may be so mildly affected that this may only become apparent on careful clinical examination. Pseudohypertrophy does not occur and the heart and intellect are unaffected. Some patients also have sensorineural deafness[55] as well as Coats' disease (a retinal disorder which may lead to retinal detachment and glaucoma). This was first reported in association with deafness and mental retardation in four sibs.[56] We have also observed a sporadic case of facioscapulohumeral dystrophy in association with retinoblastoma which was also present in one of the first reported children. The consensus from recent studies is that sensorineural deafness[57] and retinal changes[58] are both part of the facioscapulohumeral syndrome and do not delineate a separate disease entity.

Proximal muscular dystrophy

Limb girdle muscular dystrophy of adult onset is generally considered to be inherited as an autosomal recessive trait. However occasional families have been described where a clinically similar disorder is clearly inherited as an autosomal dominant trait.[59,60] In another family described in the literature[61] which the authors referred to as an 'hereditary myopathy limited to females', this was most probably an autosomal dominant trait with sex limitation. However X-linked dominant inheritance lethal in the hemizygous affected male would also be consistent with the reported pedigree. In another reported family, in which the disorder was inherited through several generations as an autosomal trait, it appears that only males were affected.[62]

Distal muscular dystrophy

In distal muscular dystrophy weakness affects the small muscles of the hands and the anterior tibial and calf muscles. It usually becomes evident in middle age. In an isolate in Sweden[63,64] two forms of the condition were recognized: (1) 'typical' cases in which only the distal muscles were affected and the disease was only slowly progressive (these cases were considered to represent the heterozygous expression of the mutant gene); and (2) 'grossly atypical' cases where the proximal muscles were also affected and the disease pursued a more rapid course (these cases were considered to represent the homozygous expression of the mutant gene among offspring of two heterozygous parents). A form predominantly affecting the distal musculature has also been described in extensive families in Britain,[65] and in North America.[66] However, in Europe and North America the condition is rare, most cases are sporadic, and it appears to be less benign than the disorder in Sweden.[67,68] It may be that the disease in Sweden is different from elsewhere. Furthermore an autosomal dominant form of distal muscular dystrophy has been described where onset is in early childhood.[69,70]

Ocular muscular dystrophy

Ocular forms of muscular dystrophy are rare and it is often difficult to establish the diagnosis since this rests upon histology and electromyography of the external ocular muscles. A limb muscle biopsy should also be performed when this diagnosis is being considered because a proportion of cases turn out to have one of the congenital myopathies (particularly myotubular and mitochondrial myopathies). Other disorders also to be excluded in the differential diagnosis include myasthenia gravis, myotonic dystrophy and various syndromes in which ophthalmoplegia may occur, such as the Kearns-Sayre syndrome.

Ocular muscular dystrophy results in ptosis and ophthalmoplegia. Onset is usually in *early* adult life.[71] The term 'oculopharyngeal' muscular dystrophy is used for a disorder in which ocular involvement is associated with progressive dysphagia and onset is usually in middle age.[72] This form of muscular dystrophy can be inherited as an autosomal dominant trait and a large proportion of such cases have been of French-Canadian origin.[73,74] A similar but much rarer condition with somewhat earlier onset has been reported among Ashkenazi Jews where it is inherited as an autosomal recessive trait.[75]

CONCLUSIONS

Under the rubric of muscular dystrophy are included some 20 or more disorders which vary considerably in their clinical features and severity. In some the disease is manifest at birth, as in the congenital forms, while in others onset may be delayed until middle age as in oculopharyngeal muscular dystrophy. They also vary considerably in their severity, some being little more than an inconvenience, as in the pure ocular muscular dystrophy, whereas others run a progressively downhill course and death may even occur as early as in the first year of life as in the rapidly progressive form of congenital muscular dystrophy. Furthermore in some forms the only tissue which seems to be affected is skeletal muscle, whereas in others, such as Duchenne muscular dystrophy, smooth and cardiac muscles are also affected as well as the brain. Finally though all these disorders are genetic they exhibit all modes of inheritance.

The only unifying feature common to all these disorders is the histological appearance of affected muscle which is characterized (to varying degrees) by variation in fibre size, fibre necrosis and phagocytosis, and ultimately replacement by fat and connective tissue. Does this shared pathology tell us anything of their underlying pathogenesis? Already it is clear that the basic biochemical defect cannot be the same because the protein dystrophin has been found to be abnormal only in the Duchenne and Becker forms of dystrophy. However, recent evidence suggests that dystrophin is associated with muscle cell membranes and may therefore be involved in maintaining the integrity of such membranes. Since the histology of affected muscle in Duchenne muscular dystrophy is at least similar to that often found in other dystrophies, could it be that in these other dystrophies the basic defect(s) resides in other membrane proteins which are also important in maintaining muscle membrane integrity? With recent developments stemming from molecular genetics and recombinant DNA technology answers to such questions may soon be found.

REFERENCES

1 Emery AEH, Dreifuss FE. Unusual type of benign X-linked muscular dystrophy. J Neurol Neurosurg Psychiatr 1966; 29: 338–342
2 Cestan R, Lejonne. Une myopathie avec retractions familiales. Nouvelle Iconographie Salpetriere 1902; 15: 38–52
3 Emery AEH. X-linked muscular dystrophy with early contractures and cardiomyopathy (Emery-Dreifuss type). Clin Genet 1987; 32: 360–367
4 Emery AEH. Duchenne muscular dystrophy. Oxford: The University Press, 1987 (revised & reprinted, 1988)
5 Voit T, Krogmann O, Lenard HG et al. Emery-Dreifuss muscular dystrophy: disease spectrum and differential diagnosis. Neuropediatrics 1988; 19: 62–71
6 Hassan Z, Fastabend CP, Mohanty PK, Isaacs ER. Atrioventricular block and supraventricular arrhythmias with X-linked muscular dystrophy. Circulation 1979; 60: 1365–1369
7 Rowland LP, Fetell M, Olarte M, Hays A, Singh N, Wanat FE. Emery-Dreifuss muscular dystrophy. Ann Neurol 1979; 5: 111–117
8 Hopkins LC, Jackson JA, Elsas LJ. Emery-Dreifuss humeroperoneal muscular dystrophy: an X-linked myopathy with unusual contractures and bradycardia. Ann Neurol 1981; 10: 230–237
9 Dickey RP, Ziter FA, Smith RA. Emery-Dreifuss muscular dystrophy. J Pediatr 1984; 104: 555–559
10 Merlini L, Granata C, Dominici P, Bonfiglioli S. Emery-Dreifuss muscular dystrophy: report of five cases in a family and review of the literature. Muscle Nerve 1986; 9: 481–485
11 Oswald A, Goldblatt J, Horak A, Beighton P. Lethal cardiac conduction defects in Emery-Dreifuss muscular dystrophy. S Afr Med J 1987; 72: 567–570
12 Hara H, Nagara H, Mawatari S, Kondo A, Sato H. Emery-Dreifuss muscular dystrophy—an autopsy case. J Neurol Sci 1987; 79: 23–31
13 Thomas PK, Calne DB, Elliott CF. X-linked scapuloperoneal syndrome. J Neurol Neurosurg Psychiatr 1972; 35: 208–215
14 Hodgson SV, Boswinkel E, Cole C, et al. A linkage study of Emery-Dreifuss muscular dystrophy. Hum Genet 1986; 74: 409–416

15 Thomas NST, Williams H, Elsas LJ, Hopkins LC, Sarfarazi M, Harper PS. Localisation of the gene for Emery-Dreifuss muscular dystrophy to the distal long arm of the X-chromosome. J Med Genet 1986; 23: 596–598

16 Yates JRW, Affara NA, Jamieson DM, et al. Emery-Dreifuss muscular dystrophy: localisation to Xq27.3—qter confirmed by linkage to the factor VIII gene. J Med Genet 1986; 23: 587–590

17 Romeo G, Roncuzzi L, Sangiorgi S, Giacanelli M, Liguori M, Tessarolo D, Rocchi M. Mapping of the Emery-Dreifuss gene through reconstruction of crossover points in two Italian pedigrees. Hum Genet 1988; 80: 59–62

18 Arahata K, Ishiura S, Ishiguro T, et al. Immunostaining of skeletal and cardiac muscle surface membrane with antibody against Duchenne muscular dystrophy peptide. Nature 1988; 333: 861–863

19 Hoffman EP, Fischbeck KH, Brown RH, et al. Characterization of dystrophin in muscle-biopsy specimens from patients with Duchenne's or Becker's muscular dystrophy. N Engl J Med 1988; 318: 1363–1368

20 Mawatari S, Katayama K. Scapuloperoneal muscular atrophy with cardiopathy. An X-linked recessive trait. Arch Neurol 1973; 28: 55–59

21 Hopkins LC, Jackson JA, Elsas LJ. Emery-Dreifuss humeroperoneal muscular dystrophy: an X-linked myopathy with unusual contractures and bradycardia. Ann Neurol 1981; 10: 230–237

22 Dickey RP, Ziter FA, Smith RA. Emery-Dreifuss muscular dystrophy. J Pediatr 1984; 104: 555–559

23 Becker PE. Dominant autosomal muscular dystrophy with early contractures and cardiomyopathy (Hauptmann-Thannhauser). Hum Genet 1986; 74: 184

24 Bailey RO, Dentinger MP, Toms ME, Hans MB. Benign muscular dystrophy with contractures: a new syndrome? Acta Neurol Scand 1986; 73: 439–443

25 Mabry CC, Roeckel IE, Munich RL, Robertson D. X-linked pseudohypertrophic muscular dystrophy with a late onset and slow progression. N Engl J Med 1965; 273: 1062–1070

26 Bergia B, Sybers HD, Butler IJ. Familial lethal cardiomyopathy with mental retardation and scapuloperoneal muscular dystrophy. J Neurol Neurosurg Psychiatr 1986; 49: 1423–1426

27 Ricker K, Mertens HG. The differential diagnosis of the myogenic (facio)-scapulo-peroneal syndrome. Eur Neurol 1968; 1: 275–307

28 Thomas PK, Schott GD, Morgan-Hughes JA. Adult onset scapuloperoneal myopathy. J Neurol Neurosurg Psychiatry 1975; 38: 1008–1015

29 Emery AEH. The nosology of the spinal muscular atrophies. J Med Genet 1971; 8: 481–495

30 Kim YJ, Chung YK, Fisk JR. Myopathy limited to the quadriceps and gastrocnemius muscles occurring in three brothers. South Med J 1979; 72: 429–432

31 McMenamin JB, Becker LE, Murphy EG. Congenital muscular dystrophy: a clinicopathologic report of 24 cases. J Pediatr 1982; 100: 692–697

32 Fukuyama Y, Kawazura M, Haruna H. A peculiar form of congenital progressive muscular dystrophy: report of 15 cases. Paediatrica (Tokyo) 1960; 4: 5–8

33 Fukuyama Y, Ohsawa M. A genetic study of the Fukuyama type congenital muscular dystrophy. Brain Develop 1984; 6: 373–390

34 Bassoe HH. Familial congenital muscular dystrophy with gonadal dysgenesis. J Clin Endocrinol 1956; 16: 1614–1621

35 Kamoshita S, Konishi Y, Segawa M, Fukuyama Y. Congenital muscular dystrophy as a disease of the central nervous system. Arch Neurol 1976; 33: 513–516

36 Lebenthal E, Shochet SB, Adam A, et al. Arthrogryposis multiplex congenita: 23 cases in an Arab kindred. Pediatrics 1970; 46: 891–899

37 Goebel HH, Lenard HG, Langenbeck U, Mehl B. A form of congenital muscular dystrophy. Brain Dev 1980; 2: 387–400

38 Furukawa T, Toyokura Y. Congenital hypotonic-sclerotic muscular dystrophy. J Med Genet 1977; 14: 426–429
39 Nonaka I, Une Y, Ishihara T, Miyoshino S, Nakashima T, Sugita H. A clinical and histological study of Ullrich's disease. Neuropediatr 1981; 12: 197–208
40 Gardner-Medwin D, Johnston HM. Severe muscular dystrophy in girls. J Neurol Sci 1984; 64: 79–87
41 Somer H, Voutilainen A, Knuutila S, Kaitila I, Rapola J, Leinonen H. Duchenne-like muscular dystrophy in two sisters with normal karyotypes: evidence for autosomal recessive inheritance. Clin Genet 1985; 28: 151–156
42 Yoshioka M, Itagaki Y, Saida K, Nishitani Y. Clinical and genetic studies of muscular dystrophy in young girls. Clin Genet 1986; 29: 137–142
43 Skyring A, McKusick VA. Clinical, genetic and electrocardiographic studies in childhood muscular dystrophy. Am J Med Sci 1961; 242: 534–547
44 Emery AEH. Abnormalies of the electrocardiogram in hereditary myopathies. J Med Genet 1972; 9: 8–12
45 Emery AEH. Hereditary myopathies. Clin Orthop 1964; 33: 164–173
46 Ben Hamida M, Fardeau M, Attia N. Severe childhood muscular dystrophy affecting both sexes and frequent in Tunisia. Muscle Nerve 1983; 6: 469–480
47 Salih MAM, Omer MIA, Bayoumi RA, Karrar O, Johnson M. Severe autosomal recessive muscular dystrophy in an extended Sudanese kindred. Dev Med Child Neurol 1983; 25: 43–52
48 Jackson CE, Strehler DA. Limb-girdle muscular dystrophy: clinical manifestations and detection of preclinical disease. Pediatrics 1968; 41: 495–502
49 Shokeir MHK, Kobrinsky NL. Autosomal recessive muscular dystrophy in Manitoba Hutterites. Clin Genet 1976; 9: 197–202
50 Shokeir MHK, Rozdilsky B. Muscular dystrophy in Saskatchewan Hutterites. Am J Med Genet 1985; 22: 487–493
51 Moser H, Wiesmann U, Richterich R, Rossi E. Progressive Muskeldystrophie VII. Haufigkeit, Klinik und Genetik der Typen I und II. Schweiz Med Wochenschr 1966; 96: 169–174, 205–211
52 Walton JN, Nattrass FJ. On the classification, natural history and treatment of the myopathies. Brain 1954; 77: 169–231
53 Walton JN, Gardner-Medwin D. The muscular dystrophies. In: Walton JN, ed. Disorders of voluntary muscle (5th edn). Edinburgh: Churchill Livingstone, 1988: p 519–568
54 Yates JRW, Emery AEH. A population study of adult onset limb-girdle muscular dystrophy. J Med Genet 1985; 22: 250–257
55 Korf BR, Bresnan MJ, Schapiro F, Sotrel A, Abroms IF. Facioscapulohumeral dystrophy presenting in infancy with facial diplegia and sensorineural deafness. Ann Neurol 1985; 17: 513–516
56 Small RG. Coats' disease and muscular dystrophy. Trans Am Acad Ophthal Otolaryn 1968; 72: 225–231
57 Voit T, Lamprecht A, Lenard HG, Goebel HH. Hearing loss in facioscapulohumeral dystrophy. Eur J Pediatr 1986; 145: 280–285
58 Fitzsimons RB, Gurwin EB, Bird AC. Retinal vascular abnormalities in facioscapulohumeral muscular dystrophy. Brain 1987; 110: 631–648
59 Chutkow JG, Heffner RR, Kramer AA, Edwards JA. Adult-onset autosomal dominant limb girdle muscular dystrophy. Ann Neurol 1986, 20: 240–248
60 Gilchrist JM, Pericak-Vance M, Silverman L, Roses AD. Clinical and genetic investigation in autosomal dominant limb girdle muscular dystrophy. Neurology 1988; 38: 5–9
61 Henson TE, Muller J, De Myer WE. Hereditary myopathy limited to females. Arch Neurol 1967; 17: 238–247
62 De Coster W, De Reuck J, Thiery E. A late autosomal dominant form of limb-girdle muscular dystrophy: a clinical, genetic and morphological study. Eur Neurol 1974; 12: 159–172

63 Welander L. Myopathia distalis tarda hereditaria: 249 examined cases in 72 pedigrees. Acta Med Scand 1951; 141 (suppl 265): 1–124

64 Welander L. Homozygous appearance of distal myopathy. Acta Genetica (Basel) 1957; 7: 321–325

65 Sumner D, Crawfurd MD'A, Harriman DGF. Distal muscular dystrophy in an English family. Brain 1971; 94: 51–60

66 Markesbery WR, Griggs RC, Leach RP, Lapham LW. Late onset hereditary distal myopathy. Neurology (Minneapolis) 1974; 24: 127–134

67 Walton JN. Clinical aspects of human muscular dystrophy. In: Bourne GH, Golarz MN, eds. Muscular dystrophy in man and animals. New York: Hafner, 1963: p 263–321

68 Kaeser HE, Wurmser P. Zum Krankheitsbild der distalen Spatmyopathie (Myopathia distalis tarda hereditaria Welander). Schweiz Med Wochenschr 1967; 97: 1208–1211

69 van der Does de Willebois AEM, Bethlem J, Meijer AEFH, Simmons AJR. Distal myopathy with onset in early infancy. Neurology (Minneapolis) 1968; 18: 383–390

70 Bautista J, Rafel E, Castilla JM, Alberca R. Hereditary distal myopathy with onset in early infancy. J Neurol Sci 1978; 37: 149–158

71 Kiloh LG, Nevin S. Progressive dystrophy of the external ocular muscles (ocular myopathy) Brain 1951; 74: 115–143

72 Bray GM, Kaarsoo M, Ross RT. Ocular myopathy with dysphagia. Neurology 1965; 15: 678–684

73 Barbeau A. The syndrome of hereditary late onset ptosis and dysphagia in French-Canada. In: Kuhn E, ed. Symposium uber progressive Muskeldystrophie. Berlin: Springer Verlag, 1966: pp. 102–109

74 Murphy SF, Drachman DB. The oculopharyngeal syndrome. J Am Med Assoc 1968; 203: 1003–1008

75 Fried K, Arlozorov A, Spira R. Autosomal recessive oculopharyngeal muscular dystrophy. J Med Genet 1975; 12: 416–418

British Medical Bulletin (1989) Vol. 45, No. 3, pp. 788–801
© The British Council 1989

Management of children: Pharmacological and physical

J Heckmatt
E Rodillo
V Dubowitz
Department of Paediatrics & Jerry Lewis Muscle Research Laboratories, Royal Postgraduate Medical School, Hammersmith Hospital, London, UK

In the absence of any effective drug treatment, physical methods of management are still the mainstay of treatment. Our current practice in Duchenne muscular dystrophy is to provide lightweight knee-ankle-foot orthoses at the time of loss of ambulation. This prolongs independent walking for an average of two years, and has the effect of delaying the onset of scoliosis, particularly if the boy remains ambulant during the pubertal growth spurt. We are currently assessing the value of radical surgery, performed early in the course of the disease, which may stabilize and prolong independent walking. In non-ambulant patients instrumentation of the spine, using mainly the Luque technique, has revolutionised the treatment of progressive scoliosis.

Ventilator support produces clinical improvement in late cases with symptomatic hypoventilation. Its place in the management of asymptomatic patients with nocturnal hypoventilation still needs evaluation, as does the role of early prophylactic respiratory support.

We have reviewed the clinical drug trials over the past 10 years. There has been an overall improvement in their quality control.

0007–1420/89/0045–0788/$10.00

In our previous contribution for the Bulletin on the management of muscular dystrophy,[1] we reviewed earlier therapeutic trials in the light of quality control, and discussed the available approaches to physical management. In the intervening decade, there have been some definite improvements in the quality of clinical trials and also some important developments in physical management. We shall discuss progress in both spheres in this review.

PHYSICAL MANAGEMENT

Use of orthoses to promote ambulation

In our previous review we discussed, in relation to the management of Duchenne muscular dystrophy, the prevention of deformities, the maintenance of posture and the promotion of ambulation using light-weight knee-ankle-foot orthoses. In this review we will discuss the use of the orthoses in the light of the latest information on the long-term results from 93 of our patients.[2] Our policy has always been to provide orthoses at the time of loss of ambulation, and not when the child is still able to walk well independently or after he has already been confined to the wheelchair.

We have published our method in detail.[3] To summarize briefly: the boy has a percutaneous Achilles tenotomy under general anaesthetic, is mobilized from the first day postoperatively in ischial weight-bearing plasters, and on the third day the plasters are removed and he is cast for the orthoses. He is then given a new set of plasters and walks in these for a week until the orthoses are ready. He spends a further week to ten days in hospital until he has achieved stable walking in the orthoses.

Regular assessment while the boy is still walking independently allows us to predict when he will need intervention and to familiarize the family with our philosophy of rehabilitation. It also allows detection of the occasional unsuitable child either because of adverse psychological factors, or the presence of marked asymmetry of hip and knee contractures. Average duration of walking in the orthoses for the 93 consecutive cases reviewed was 21 months, including 7 early in the program who did not walk at all due to ill fitting orthoses. The longest time was 72 months.

Walking in orthoses gives the child independence, is of psychological benefit, and prevents the development of scoliosis. In our recent survey, scoliosis did not usually develop until the boy lost ambulation in the orthoses. Once he was in the wheelchair the rate of deterioration of the scoliosis related to his age. It was rapid ($2.4°/$ year) between the ages of 13–15 years, i.e. the pubertal growth spurt, as against less than $1°/$ year the rest of the time. Rapid deterioration of the scoliosis was prevented in the 22% of boys who walked during puberty in their orthoses. Those who stopped walking in the orthoses before puberty, still developed scoliosis during puberty but the orthoses postponed its development by an average of two years.[2]

Surgical intervention

Late

Rideau et al.[4] claimed a two year extension of independent non-braced ambulation in 10 boys with Duchenne muscular dystrophy by the use of radical surgery. The technique involved: (1) correction of Achilles tendon contractures and pes cavus; (2) transfer of the tibialis posterior muscle to the dorsum of the foot; (3) extensive tenotomies of contracted muscles around the hip, especially the ilio-tibial tract; and (4) occasional transplant of the distal insertion of the iliopsoas muscle to correct the forward pelvic tilt. The patients had one week of bedrest post-operatively, followed by 2 weeks physiotherapy in the swimming pool and then three weeks learning non-assisted ambulation. The authors also treated a group of 10 patients with other forms of muscular dystrophy in the same way.

Early

Rideau and co-workers have subsequently modified their method to treat younger Duchenne patients, between the ages of 4–6 years. The surgery was tailored to the individual child, and usually involved lengthening of the achilles tendons, removal of the fascia lata by a long incision down the side of the thigh, and transection of the rectus femoris tendon at the hip. Percutaneous release of other tight structures, such as the hamstring tendons behind the knee, might also be required.

These workers claimed that such early surgery gave the child a normal gait and freedom from regular physiotherapy, and either stabilized or improved his function. Their method of treatment demands a re-think of the existing philosophy of the management of Duchenne muscular dystrophy, which has been to advise surgical intervention only at the time of providing orthoses, and to avoid any radical correction of deformities as this may precipitate loss of ambulation if performed in the later stages of independent ambulation. The long term result of early radical surgery is still unknown, and we are currently undertaking a randomized controlled trial of the procedure.

Management of scoliosis

The Luque operation for the prevention of progressive scoliosis[5] has been the most significant therapeutic advance for this severe complication, which occurs in some 90% of non-ambulant Duchenne patients. External supports such as the spinal brace and molded seat are not effective in preventing scoliosis; it is not even certain that they appreciably slow down the rate of progression. Their use may have the adverse effect of delaying operative fixation of the spine until the child already has significant respiratory muscle weakness.[6,7]

The Luque operation involves stabilization of the spine by means of two internal rods and laminar wiring to individual vertebrae. The laminar wiring applies a corrective force at many points along the spine, rather than just at the ends of the rod as with the earlier Harrington technique. Luque has emphasized that the operation provides fixation of the spine in inherently unstable situations, such as the soft spinal bone of paralytic scoliosis.[5] This stability allows early mobilization of the patient on the second day post-operatively without the need for external support of a jacket, and there is less likelihood of later failure. In contrast, the Harrington technique, which relies on a bone graft for final stability of the spine, requires the child to be immobile for a week postoperatively and to use a spinal brace for some months. The Luque rod can be moulded to a wide variety of deformities, and in particular can allow some thoracic kyphosis, which is important for arm function and head posture in Duchenne patients, while the Harrington rod tends to extend the thoracic spine.[8,9]

The major contraindication to surgery is a vital capacity below 30% of expected. A number of reports have highlighted the risk of post-operative respiratory complications in such patients. This has led to the recommendation for early surgery while the patient has a mild 30° curve and evidence of progression.[9-12] Rideau et al.[13] have emphasized that a subgroup of patients who lose ambulation before the age of 10 years are likely to have restriction of vital capacity below the critical 30% before a 30° curve develops, and therefore should have early surgery, in anticipation of a curvature developing, at a stage when the spine is still straight.

The long-term results of early Luque instrumentation in Duchenne muscular dystrophy are still awaited. The spine may continue to grow and the deformity increase during puberty, necessitating a second operation which would clearly be a major undertaking. Methods which prolong ambulation and postpone the onset of scoliosis, such as the use of orthoses or early radical surgery, would therefore seem worthwhile. The vertebrae will be better ossified and better able to take the rod, and there will be less post-operative growth.

Respiratory care

Progressive respiratory muscle weakness is the cause of death in 70–80% of patients with Duchenne muscular dystrophy.[14,15] Cardiac causes are responsible for the remainder, and in some cases it is possible that respiratory failure may contribute to the cardiac involvement.[16] Diaphragm weakness, an important cause of respiratory failure in neuromuscular disorders, occurs late in Duchenne muscular dystrophy and carbon dioxide retention is a terminal event.[14,17,18]

A number of workers have provided mechanical ventilatory support for non-ambulatory Duchenne patients in respiratory failure. These include the rocking bed, the cuirass pump, the iron lung, intermittent positive pressure mouth ventilation, and positive pressure ventilation via tracheostomy. Patients used the ventilatory assistance at night, and some required occasional assistance during the day. All authors reported relief of distressing symptoms, such as insomnia, progressive drowsiness, morning headaches and marked anxiety, and claimed extended survival for periods up to 10 years.[15,19–22]

Despite these claims, none of these methods of ventilation are really satisfactory. Much of the equipment is expensive, complex and cumbersome. It restricts mobility, causes increasing isolation, and increases the need for overnight care.[23,24] Furthermore, from the mechanical point of view these methods are not always effective. While positive pressure ventilation via tracheostomy is effective, it has the ethical problem of being invasive and irreversible.

Intermittent positive pressure ventilation with a nasal mask is an important recent advance which may have useful application to Duchenne muscular dystrophy.[25] It seems to be effective mechanically, the equipment is portable and relatively quiet in operation, the mask is easy to apply, and considerable variation of sleeping posture is possible without leakage. The equipment is more reasonably priced than the iron lung or rocking bed. A potential advantage of ethical importance is that prolongation of life beyond the point of incapacitating bulbar weakness is less likely than with tracheostomy.[26]

The frequency of asymptomatic sleep hypoventilation in Duchenne patients has recently been demonstrated by Smith et al.[27] who found severe oxygen desaturation during REM sleep in nine out of 14 cases, with an average age of 18 years, studied overnight with polysomnography. None of the patients had symptoms, presumably because of the restorative effect of non-REM sleep. In the light of these findings, intervention at the asymptomatic stage may be worthwhile. Although there are several practical and ethical issues involved, one might argue that it is paradoxical to perform the substantial Luque operation to stabilise the spine, and not provide prophylactic nasal ventilation. Only a prospective controlled trial of prophylaxis, can determine potential long-term relief of symptoms and extension of useful life.

THERAPEUTIC TRIALS

At the time of our earlier review,[1] there was much to criticize about the state of clinical trials in muscular dystrophy. In several instances benefit had been claimed for a drug following an uncontrolled trial of poor quality, necessitating a repeat trial with proper controls to prove that the drug was of no value. Although most recent trials are of good quality, problems of methodology and controversies over trial design remain. These controversies relate particularly to the method of assessment, the age and functional range of the patients studied, and what type of trial offers the most efficient means of drug evaluation.

One particular feature of recent trials has been the better definition of cases and the concentration on a fairly pure cohort of Duchenne patients. Although our remarks below relate mainly to therapeutic studies in Duchenne muscular dystrophy, they are also relevant to the other muscular dystrophies.

When considering the assessment of the drug response, a measurement which is 'well behaved' and changes linearly with time is relatively easy to analyze. A measurement which shows rapid deterioration during some part of the illness, is very difficult to analyze. Muscle strength is a well behaved measurement whereas timed tasks and functional scores are not.[28-30] It has been claimed that badly behaved measurements may be converted into standard deviation scores of the normal range, in which case they are more suitable for analysis.[31] It is difficult to see how such transformation can work unless the normal range also changes in a similar non-linear fashion.

The main value of functional assessment is to detect potential differences in disease severity in the treatment and control groups at entry into the trial, as this influences subsequent clinical progress (see below).[30] For this purpose, the 40 point motor ability score[32] gives better definition than the original 10 point Vignos scale.

The power of a clinical trial is determined by the size of the standard deviation in relation to the mean of whatever measurement has been performed, which in most trials of Duchenne muscular dystrophy is the deterioration in muscle strength. The less the power of the trial the larger the number of patients required. On the 10 point MRC scale, Brooke et al.[29] reported rather a low power. This led to the need for large multicentre trials.[33,34] The main reason for low power was probably the inclusion of patients from a wide age range. Inclusion of non-ambulant patients means reliance on measurement of distal muscles which show a relatively small change with time.[35] Inclusion of patients under five years means that changes in muscle strength are difficult to measure because of problems with cooperation, while the child's functional ability may improve.

Various strategies have been adopted to overcome the difficulties produced by including non-ambulant patients. Dick et al.[36] and Bertorini et al.[37] used matched pairing techniques based on functional activity and age. Moxley et al.[38] defined a subset of patients who were ambulant. None of these reports gave sufficient details of methodology, and in two the technique was retrospective, which raises doubts about objectivity. It is also important that two of the trials were small[36,37] and only about half the patients were independently ambulant at entry.

In our previous review we advocated a two year trial period as a minimum for assessing treatment response in Duchenne muscular dystrophy, because the pattern of progression is variable and there may be periods of apparent clinical arrest.[1] A further reason is that sequential measurements of muscle strength show fluctuations which are not motivational but relate to physiological changes.[35] These fluctuations are small in absolute terms (i.e. in relation to normal strength at the same age), but significant relative to the patient's weakness. A trial period of one year is reasonable, but only if all patients are ambulant throughout.[30]

If every drug requires a multicentre trial for evaluation, this will become a recipe for inertia and a threat to trying out new ideas.[39] This is not to deny the contribution this type of study can make to accurate drug evaluation, but a symptom of the problems it introduces is the recent recommendation for preliminary trials with historical controls.[31,40] Edwards et al.[39] have argued for short-term prospective double-blind studies involving combined clinical and biochemical analysis. A similar approach was the double-blind Mazindol trial by Zatz et al.[41] on a pair of ambulant identical twins. A third approach is the unicentre prospective double-blind randomized controlled trial performed on a relatively small number of ambulant patients.[30]

There are many problems with the use of historical controls. The inclusion criteria for the treatment and control groups may vary, there may be a change in the experimental environment, and in the subjective evaluation during the prolonged period of observation.[42,43] In claiming a beneficial effect of prednisolone, Brooke et al.[40] could not avoid the other defect of such studies, which is placebo bias. They claimed not to have seen a placebo effect in their previous trials, but this is at variance with our experience.[30] The short duration of their trial, only six months, would have increased the likelihood of a placebo effect interfering with the trial result.

Selective bias is a major problem of trials which use historical controls. This criticism applies to the prednisolone trial of De Silva et al.[44] in which the main criterion used for response to the drug was a comparison of the age at which the child went into the wheel-chair. This criterion is influenced by the child's functional status at the start of the trial,[30] and the authors did not show that their two groups of patients were functionally comparable at entry into the trial. By definition boys with Duchenne muscular dystrophy lose ambulation by their 13th birthday, but this can vary in severity from 6 to 12 years of age.

If one can show a drug does not produce the expected biochemical effect, then the theoretical basis for giving the drug is in doubt. For example, Griffiths et al.[45] showed that neither allopurinol nor

Table 1   Results of therapeutic trials using various substances

Reference	Drugs & other substances	Patients treated	Ages	Controls used	Ages	Methods of evaluation	Treatment value	Quality score
1. Amino acids								
Mendell et al.[33]	Leucine	47 (DMD)	9.2 years (? range)	44 (DMD)	9.8 years	10 point MRC scale Timed functional tasks etc[29]	No value	5
2. Steroids								
Brooke et al.[40]	Prednisolone 1.5 mg/kg/day	33 (DMD)	5–15 years	170 (DMD) Historical controls[31]	2–20 year	As Brooke et al[29]	Definite improvement.	2
De Silva et al.[44]	Prednisolone 2 mg/kg/day	20 (DMD)	3–16 years	38 (DMD) Historical controls	0.1–8 years	Age loss ambulation	Long term benefit	1
3. Allopurinol								
Hunter et al.[46]	Allopurinol 100 mg/day	21 (DMD)	3–16 years	Cross over, 3 months each limb	—	Grip strength MRC score Timed functions	No effect	4
Bertorini et al.[47]	Allopurinol 300 mg/day Adenine 300 mg day	7 (DMD)	8–15 years	7	8–17 years	As Brooke et al.[29] Open biopsy	No value	4

Calcium channel blockers							
Dick et al.[36]	Flunarizine 0.1–0.25 mg/kg/day	13 (DMD)	5–14	13 (DMD)	5–14	MRC, Timed tasks, locomotor score	no value	5
Moxley et al.[38]	Nifedipine 0.75–2 mg/kg/day	52 (DMD) 2 (BMD)	2–27	50 (DMD) 1 (BMD)	—	As Brooke et al.[29]	no value	5
Bertorini et al.[37]	Diltiazem 8 mg/kg/day	11 (DMD)	6–16 years	11 (DMD)	6–18 years	As Brooke et al.[29] Needle biopsy	no value	4
5. Other drugs							
Patten & Zeller[48]	Methysergide 8 mg/day	8	6–8 years	Cross over 3 months each limb	—	Timed functions, grip strength	no value	3
Heckmatt et al[30]	Isaxonine 25 mg/kg/day	9 (DMD)	6–10 years	10 (DMD)	6–8 years	Myometry, MRC, timed walking, locomotor score[32]	no value	6

ribose influenced the ATP state of the muscle as assessed non-invasively by ^{31}P-NMR spectroscopy. Even repeated biochemical measurements of muscle by biopsy may be possible in selected patients using the needle technique. Combining a biochemical assessment with a short-term (6–12 months) double-blind clinical assessment in a few patients allows rapid screening of a number of drugs without embarking on a large scale trial. This type of fundamental approach may be attractive to drug companies who are more interested in posing basic scientific questions of potential application to several diseases, rather than trying empirically to develop a drug for a single disease such as muscular dystrophy which is comparatively rare.[39] When patient numbers are small, this approach is only valid when the activity of a particular drug is known and can be easily measured. It will become more applicable, however, with progressive understanding of the fundamental defect in Duchenne muscular dystrophy.

The controlled double-blind unicentre trial on ambulant patients with accurate clinical assessments of muscle strength and other parameters gives a trial of acceptable power on relatively few patients. Our Isaxonine trial which followed these principles proved to have greater statistical power than the large multicentre trials.[30]

In Table 1, we have listed the therapeutic trials in Duchenne muscular dystrophy published since our last review. Three of these have not been discussed above.[46-48] For consistency we have continued to use the same quality control score. These were: (1) careful selection and definition of cases; (2) use of adequate controls; (3) whether or not the trial was performed blind; (4) assessment of patients by other than simple clinical ratings; and (5) a trial period of at least 2 years. We have added an additional score (6) for prospective trials which were shown by the authors to have an 80% chance of detecting a 75% slowing of the disease process by the drug on a one-tail test.

In the scoring system, we have allowed a point for patient selection if non-ambulant cases were included, so long as all the patients had classical Duchenne muscular dystrophy. We have not scored for the use of historical controls. We have scored prospective double-blind controlled trials even when patient numbers were small. It might be argued that small trials should not receive any points, but we think that if the design of such trials is fundamentally scientific they are potentially more useful than large poorly controlled studies. We have scored a carefully validated

MRC scoring system, but not measurement of grip strength, however accurate. Our own study[30] scored 6 because we designed it to comply with our previous recommendations. Others may wish to recommend a quality score which satisfies different criteria.

The recent dramatic advances in relation to the isolation of the Duchenne gene and the discovery of dystrophin, the protein it encodes, have raised hopes for the potential treatment of Duchenne dystrophy by gene or protein replacement at the early stage of the disease. Studies on the animal models, the *mdx* mouse and the X-linked canine dystrophy, may provide an alternative to human trials but will need equally stringent controls and accurate quantitation of muscle function. Recent studies on the implantation of normal precursor muscle cells directly into normal or dystrophic muscle in mice have shown that donor muscle cells can proliferate in the host muscle, and in addition hybridize with host myoblasts to produce dimeric fibres.[49-51] If such an approach were clinically practical it could provide a universal form of treatment for different forms of X-linked recessive or autosomal recessive dystrophy, without knowledge of the underlying gene or fundamental defect. The scene is now set for applying new approaches to therapy for the human subject, once the technical and immunological difficulties have been overcome in the animal models. Meanwhile, supportive therapy remains the priority in maintaining optimal function in these children until the new breakthrough becomes a reality.

REFERENCES

1 Dubowitz V, Heckmatt JZ. Management of muscular dystrophy—pharmacological and physical aspects. Br Med Bull 1980, 36, 139–144
2 Rodillo EB, Fernandez-Bermejo E, Heckmatt JZ, Dubowitz V. Prevention of rapidly progressive scoliosis in Duchenne muscular dystrophy by prolongation of walking with orthoses. J Child Neurol, 1988 (in press)
3 Heckmatt JZ, Dubowitz V, Hyde SA, Florence J, Gabain A, Thompson N. Prolongation of walking with light-weight orthoses: review of 57 cases. Dev Med Child Neurol 1985; 27: 149–154
4 Rideau Y, Glorion B, Duport G. Prolongation of ambulation in the muscular dystrophies. Acta Neurol (Bern) 1983; 5: 390–397
5 Luque ER. The anatomical basis and development of segmentation of spinal instrumentation. Spine 1982; 7: 256–259
6 Robin GC. Scoliosis in Duchenne muscular dystrophy. Israel J Med Sci 1977; 13: 203–206
7 Moseley CF. Natural history and management of scoliosis in DMD. In Serratrice et al, Neuromuscular diseases. New York: Raven Press, 1984
8 Sullivan JA, Conner SB. Comparison of Harrington instrumentation and segmental spinal instrumentation in the management of neuromuscular spinal deformity. Spine 1982; 7: 299–304

9 Sussman MD. Advantage of early spinal stabilization and fusion in patients with Duchenne muscular dystrophy. J Pediatr Orthop 1984; 4: 532–537

10 Jenkins JG, Bohn D, Edmonds JF, Levison H, Barker GA. Evaluation of pulmonary function in muscular dystrophy patients requiring spinal surgery. Crit Care Med 1982; 10: 645–649

11 Kumano K, Tsuyama N. Pulmonary function before and surgical correction of scoliosis. J Bone Joint Surg 1982; 64A: 242–248

12 Cambridge W, Drennan JL. Scoliosis associated with Duchenne muscular dystrophy. J Pediatr Orthop 1987; 7: 436–440

13 Rideau Y, Glorion B, Delaubier A, Tarle O, Bach J. The treatment of scoliosis in Duchenne muscular dystrophy. Muscle Nerve 1984; 7: 281–286

14 Inkley SR, Oldenberg FC, Vignos PJ. Pulmonary function in Duchenne muscular dystrophy related to stage of disease. Am J Med 1974; 56: 297–306

15 Rideau Y, Gatin G, Bach J, Gines G. Prolongation of life in Duchenne's muscular dystrophy. Acta Neurologica 1983; 38: 118–124

16 Smith PEM, Calverly PMA, Edwards RHT, Evans GA, Campbell EJM. Practical problems in the respiratory care of patients with muscular dystrophy. New Engl J Med 1987; 316: 1197–1204

17 Newsom-Davis J, Goldman M, Loh L, Casson M. Diaphragm function and alveolar hypoventilation. Q J Med 1976; 45: 87–100

18 Newsom-Davis J. The respiratory system in muscular dystrophy. Br Med Bull 1980; 36: 135–8

19 Alexander MA, Johnson EW, Petty J, Stauch D. Mechanical ventilation of patients with late stage Duchenne muscular dystrophy: management in the home. Arch Phys Med Rehabil 1979; 60: 289–292

20 Bach J, Alba A, Pilkington LA, Lee M. Long-term rehabilitation in advanced stage of childhood onset, rapidly progressive muscular dystrophy. Arch Phys Med Rehabil 1981; 62: 328–331

21 Curran FJ. Night ventilation by body respirators for patients in chronic respiratory failure due to late stage Duchenne muscular dystrophy. Arch Phys Med Rehabil 1981; 62: 270–274

22 Bach JR, O'Brien J, Krotenberg R, Alba AS. Management of end stage respiratory failure in Duchenne muscular dystrophy. Muscle Nerve 1987; 10: 177–182

23 Colbert AP, Schock NC, Respirator use in progressive neuromuscular disease. Arch Phys Med Rehabil 1985; 56: 760–762

24 Miller JR, Colbert AP, Schock NC. Ventilator use in progressive neuromuscular disease: impact on patients and their families. Dev Med Child Neurol 1988; 30: 200–207

25 Ellis ER, Bye PT, Bruderer JW, et al. Treatment of respiratory failure during sleep in patients with neuromuscular disease: positive-pressure ventilation through a nose mask. Am Rev Respir Dis 1987; 135: 148–152

26 Heckmatt JZ. Respiratory care in muscular dystrophy. (Editorial). Br Med J 1987; 295: 1014

27 Smith PEM, Calverly PMA, Edwards RHT. Hypoxia during sleep in Duchenne muscular dystrophy. Am Rev Respir Dis 1988; 137: 884–888

28 Brooke MH, Griggs RC, Mendell JR, Fenichel GM, Shumate JB, Pellegrino RJ. Clinical trial in Duchenne dystrophy. 1. The design of the protocol. Muscle Nerve 1981; 4: 186–97

29 Brooke MH, Fenichel GM, Griggs RC, et al. Clinical investigation in Duchenne muscular dystrophy: 2. Determination of the 'power' of therapeutic trials based on the natural history. Muscle Nerve 1983; 6: 91–103

30 Heckmatt JZ, Hyde SA, Gabain A, Dubowitz V. Therapeutic trial of isaxonine in Duchenne muscular dystrophy. Muscle Nerve 1988; 11: 836–847

31 Mendell JR, Province MA, Moxley RT, et al. Clinical investigation of

Duchenne muscular dystrophy. A methodology for therapeutic trials based on natural history. Arch Neurol 1987; 44: 808–11

32 Scott OM, Hyde SA, Goddard C, Dubowitz V. Quantitation of muscle function in children: a prospective study in Duchenne muscular dystrophy. Muscle Nerve 1982; 5: 291–301

33 Mendell JR, Griggs RC, Moxley RT, et al. Clinical investigation of Duchenne muscular dystrophy: 4. Double blind controlled trial of leucine. Muscle Nerve 1984; 7: 535–41

34 Moxley RT, Brooke MH, Fenichel GM, et al. Clinical investigation of Duchenne muscular dystrophy: 6. Double-blind controlled trial of nifedipine. Muscle Nerve 1985; 10: 22–33

35 Edwards RHT, Chapman SJ, Newham DJ, Jones DA. Practical analysis of variability of muscle function measurements in Duchenne muscular dystrophy. Muscle Nerve 1987; 10: 6–14

36 Dick DJ, Gardner-Medwin D, Gates PG, Gibson M, Simpson JM, Walls TJ. A trial of flunarizine in the treatment of Duchenne muscular dystrophy. Muscle Nerve. 1986; 9: 349–54

37 Bertorini TE, Palmeri GMA, Griffin JW, et al. Effect of chronic treatment with the calcium antagonist diltiazem in Duchenne muscular dystrophy. Neurology 1988; 38: 609–13

38 Moxley RT, Brooke MH, Fenichel GM, et al. Clinical investigation in Duchenne dystrophy. VI. Double-blind controlled trial of nifedipine. Muscle Nerve 1987; 10: 22–33

39 Edwards RHT, Jones DA, Jackson MJ. An approach to treatment trials in muscular dystrophy with particular reference to agents influencing free radical damage. Med Biol 1981; 62: 143–7

40 Brooke MH, Fenichel GM, Griggs RC, et al. Clinical investigation of Duchenne muscular dystrophy. Interesting results in a trial of prednisolone. Arch Neurol 1987; 44: 812–817

41 Zatz M, Betti RTB, Frota-Pessoa O. Treatment of Duchenne muscular dystrophy with growth hormone inhibitors. Am J Med Genet 1986; 24: 549–566

42 Pocock SJ. Clinical trials, a practical approach. Chichester: Wiley 1983, pp. 54–60

43 Peto R, Pike MC, Armitage P, et al. Design and analysis of randomised clinical trials requiring prolonged observation of each patient. Br J Cancer 1976; 34: 585–612

44 De Silva S, Drachman DB, Mellits D, Kunel RW. Prednisolone treatment in Duchenne muscular dystrophy. Arch Neurol 1987; 44: 818–822

45 Griffiths RD, Cady EB, Edwards RHT, Wilkie DR. Muscle energy metabolism in Duchenne muscular dystrophy studied by ^{31}P-NMR: controlled trials show no effect of allopurinol or ribose. Muscle Nerve 1985; 8: 760–767

46 Hunter JR, Galloway JR, Brooke MH, et al. Effects of allopurinol in Duchenne muscular dystrophy. Arch Neurol 1983; 40: 1294–1299

47 Bertorini TE, Palmieri GM, Griffin J et al. Chronic allopurinol and adenine therapy in Duchenne muscular dystrophy: effects on muscle function, nucleotide degradation, and muscle ATP and ADP content. Neurology 1985; 35: 61–65

48 Patten BM, Zeller RS. Clinical trials of vasoactive and anti serotonin drugs in Duchenne muscular dystrophy. Ann Clin Res 1983; 15: 164–165

49 Watt DJ, Morgan JE, Partridge TA. Use of mononuclear precursor cells to insert allogenic genes into growing mouse muscles. Muscle Nerve 1984; 7: 741–750

50 Law PK, Goodwin TA, Wang MG. Normal myoblast injections provide genetic treatment for murine dystrophy. Muscle Nerve 1988; 11: 525–533

51 Partridge TA, Morgan JE, Coulton GR, Hoffman EP, Kunkel LM. Conversion of mdx myofibres from dystrophin-negative to -positive by injection of normal myoblasts. Nature 1989; 337: 176–179

British Medical Bulletin (1989) Vol. 45, No. 3, pp. 802–818
© The British Council 1989

Management of muscular dystrophy in adults*

Richard H T Edwards
University of Liverpool Muscle Research Centre, Royal Liverpool Hospital,
Liverpool, UK

Muscular dystrophies in adults present a spectrum of
disabilities and medical problems which require diagnosis
and timely management. Weight control and orthotic
assistance can help the sufferer maintain as near a normal
life style as possible. Surgery may offer benefits in selected
patients. Major issues such as employment, personal/sexual
relations, parenthood, chronic disability and handicap,
require clear analysis and sympathetic management. Later,
respiratory and/or cardiac failure become significant
considerations. Currently the move is towards a policy of
respiratory support but it is not yet clear how much this may
improve the quality or length of life.

The muscular dystrophies seen in adult life (Table 1) present
many challenges, not least in establishing the diagnosis. The
diagnosis of a muscular dystrophy[1] is based on taking a medical
(including family) history and careful clinical examination which
is still of cardinal importance despite modern diagnostic techniques.[2] Percutaneous muscle biopsy with a 'needle'[3] or concho-
tome[4] is a practical method of obtaining muscle samples for
histopathology and histochemistry. This paper seeks to present
the principles of management in the recognition that the patient's
needs are more related to the degree of impairment, disability or
handicap (Table 2) than to the specific diagnosis.

*Based in part on the Keynote Address 'General developments in research and
treatment in neuromuscular diseases. The multidisciplinary approach'. European
Alliance of Muscular Dystrophy Associations (EAMDA): 17th European Conference, Oslo September 1987. Published by Foreningen for Muskelsyke. Sporveisgata 10, N-0354 Oslo 3, Norway 1988.

0007–1420/89/0045–0802/$10.00

Table 1   Muscular dystrophy in adults (most usual presentations at adult muscle clinics)

Inheritance	Diagnosis	Onset
X-LINKED (short arm X-chromosome (same gene) (long arm X-chromosome	Duchenne dystrophy Becker dystrophy Emery-Dreifuss	3–5 years but presentation as severely disabled childhood childhood
AUTOSOMAL RECESSIVE	Scapulohumeral (Limb girdle)* Congenital	early adult life early adult life hypotonia at birth
AUTOSOMAL DOMINANT	Facioscapulohumeral Scapuloperoneal Oculopharyngeal Distal	any age second-fifth decade adulthood adulthood
AUTOSOMAL DOMINANT (Chromosome 19)	Myotonic dystrophy	adulthood usually

*Limb Girdle Dystrophy is a diagnosis which is made less frequently as improved diagnostic facilities point to scapulohumeral or Becker muscular dystrophy.

Table 2   Muscular dystrophy some milestones for the patient and family

Problem	Components	Management options
EARLY Impairment	'Suspicion' 'Diagnosis' 'Prognosis' Genetic risks	Correct diagnosis Encouragement/support Personal responsibility Counselling
Disability	Muscle weakness Walking difficulties Risks of falling Loss of ambulation	Orthotics Weight control Corrective surgery -limbs-spine -scapular fixation
Handicap	Employment problems Mobility Sexual relations	Education Expert advice Counselling Patient support groups
LATE	Loss of independence respiratory problems The final illness Bereavement	Sheltered accommodation Helpers Confidence in local Medical support Counselling Patient support groups Voluntary work on behalf of others

INTEGRATED APPROACH TO CARE

This integrated approach is based in practice on several clinics for children and adults in Liverpool and a combined clinic (Metabolism, Orthopaedics, Paediatrics and Clinical Genetics) which comprises a wide range of paramedical professional supports (Physiotherapy, Dietetics, Orthotics, Occupational Therapy) held at the Robert Jones and Agnes Hunt Orthopaedic Hospital, Oswestry, Shropshire some 50 miles (80 km) south of Liverpool.

Our attitude to the patients is echoed in the message 'Everyone's different, nobody's perfect'[5] and every encouragement is given to achieve as much independence as possible. Literature is now available to give guidance[6-11] to sufferers and their families on how they can help in the management of problems encountered (Table 2). A welcome development is the appointment by the Muscular Dystrophy Group of Great Britain and Northern Ireland of Family Care Officers (FCO's) to help in advising patients and their families by visiting them in their homes, thus forming a 'bridge' with the care provided by the clinics. It is particularly helpful to have the FCO's report of a home visit written in the case notes for review and discussion with the patient when seen in the clinic. These paramedical professional staff help provide such essential information as how to obtain benefits from the Social Services and where to obtain standing frames, wheel chairs etc, as well as providing much emotional support. Good physical care is emphasized to prevent secondary complications leading to deformities which early influence posture and the ability to walk, and later (and more seriously) respiration. Formal gait analysis may be a valuable indicator of the progress of the secondary consequences of contractures of hips, knees and ankles.[12,13]

Weight control

Weight reduction by dieting has been shown to be metabolically safe.[14] This is of paramount importance for the preservation of mobility because of the obvious effect of improving the power/-weight ratio. For Duchenne muscular dystrophy it has been possible to make estimates of how the progressive loss of muscle influences the growth curves so as to help define target weights at different ages.[15,16] It is important in chair- or bed-bound patients to keep the weight down to avoid obesity-related respiratory complications and to make nursing care easier. This advice may need to be modified if pressure areas develop over the supporting promontories (e.g. the sacrum of greater trochanters).

PHYSIOTHERAPY

The common sense principles of physiotherapy in these diseases are care of posture, prevention of contractures, maintenance of muscle strength and preservation of respiration by instruction of relatives in the techniques of postural drainage and assisted coughing. In the childhood dystrophies there is specific management with exercises and properly designed and fitting orthoses.[12] In adults who have stopped growing there is less to be gained by stretching other than by the use of a standing frame. It does seem helpful however to move as many joints as practical through their range of movement regularly to ease the patient's discomfort from being immobilized for many hours. Swimming is an enjoyable recreation and a means of maintaining joint mobility and strength without overloading the muscles. Whether the course of the disease is altered by exercise is still a matter for discussion.[17,18]

If the pharyngeal or expiratory muscles are particularly affected there may be more specific requirements such as care of a nasogastric tube to prevent risk of aspiration which is however a greater problem when weakness is due to motor neurone disease.[19,20] This is mentioned because aspiration has been a cause of death in a teenager with Duchenne muscular dystrophy.

ORTHOPAEDIC SURGERY

Here the underlying principle is to have as short a period of immobilization as possible. Fortunately modern orthopaedic surgery has developed a number of effective techniques for internal fixation. A general consequence of corrective surgery is that useful 'trick' movements may be lost with an immediate deterioration in motor performance but generally others can be learned so that with correction of deformity improved performance may reasonably be expected.

For contractures in the lower limbs

Occasionally (i.e. less than in childhood dystrophy) it can be helpful to consider referral for release of muscle shortening, tendon transfer or other procedures to help control muscle forces acting on a weight bearing joint such as the knee or ankle.[12] Great care is needed in the preoperative assessment lest the operation have the undesirable effect of destabilizing the knee thus preventing future standing or walking.[13]

For upper limb girdle weakness

Weakness of the proximal muscles of the shoulder girdle as in facioscapulohumeral muscular dystrophy is profoundly disabling even if most of the arm muscles retain strength. Surgical fixation of the scapula to the thorax by screw attachment (Fig. 1), taping or wiring to the ribs together with provision for later bony union can help preserve the upper limb function.[12] In the preoperative assessment it is our practice to use a bimanual grip to hold the scapula stable while the patient attempts to raise the arm above the head. If the patient can do so ('the first time for years' is a comment not infrequently heard) then the operation may be justified. If weakness has extended down the arm to a degree that no obvious improvement in performance is achieved by this manoeuvre then surgery is unlikely to be helpful.

For spinal deformity

Control of spinal curvature in muscular dystrophy by the recently introduced (Luque) method of internal fixation offers great advantages[21-23] in Duchenne muscular dystrophy but also poses a difficult decision for the patient and his family. By now the accumulated experience from the diaries kept over the period indicate that the teenagers are pleased with the result because of improved head control, posture and appearance. These impressions are clearly conveyed by the patient who has undergone the Luque procedure and this can help to give encouragement to those making the decision.

It is not common to refer adults with muscular dystrophies for spinal surgery except to relieve pain or instability. A gait analysis[13] done with and without a spinal brace can give some indication of the possible consequence of spinal fixation. In adults with profound proximal muscle weakness it is evident that walking is only possible because the pelvis is raised as a result of 'throwing' the body to the opposite side giving the appearance of supranormal mobility of the lumbar spine. Spinal fixation may not only prevent

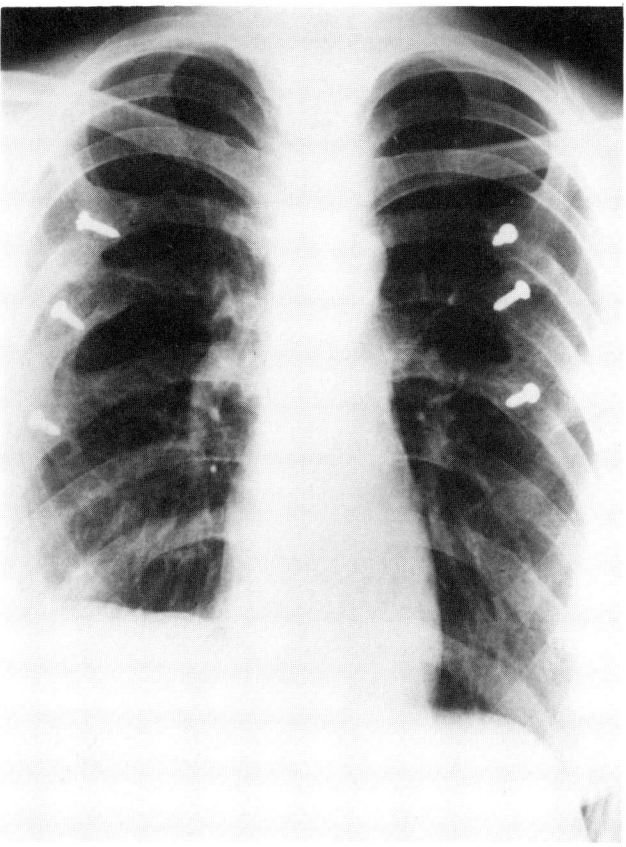

Fig. 1 Chest radiograph of a female patient aged 35 with facioscapulohumeral muscular dystrophy. She underwent bilateral scapular fixation 16–18 years previously with good retention of use of arms (she could still raise her arms above her head). Note the screws used to fix the scapulae to the ribs before bony union had been achieved and the elevated right diaphragm indicating inspiratory muscle weakness.

this useful adaptation but also prevent ambulation by destabilizing the knee joint. It is not surprising therefore that a patient may refuse an operation if wearing a spinal plaster jacket or brace disturbs his or her standing posture or balance. Severe lumbar lordosis in the chairbound adult with severe muscular dystrophy can also present problems of back pain and the unusual risk of paraplegia with loss of bowel and bladder control.

PLASTIC SURGERY

To correct for failure of eye closure

In facioscapulohumeral muscular dystrophy weakness of eye closure is common and drying of the sclerae can result in conjunctivitis or worse, ulceration. It may be sufficient to use 'artificial tears' but if these do not control the problem it may be necessary to refer the patient for tarsorrhaphy.

To correct for failure of mouth closure

In facioscapulohumeral muscular dystrophy severe facial weakness may be associated with a disfiguring and inconvenient paralysis of the muscles of the mouth. Referral for plastic surgery can give enormous benefit in restoring self respect from the improved facial appearance and the ability to eat without slobbering.

OTHER CONSEQUENCES OF MUSCLE WEAKNESS

Abdominal muscle weakness contributes to difficulty in bowel action, consequently constipation is a common problem which has to be managed by standard methods to minimize distress and the risk of distension embarassing breathing. The extent to which weakness of the smooth muscle of the bowel wall also contributes to constipation is uncertain in individual cases but is well recognized in myotonic dystrophy[24,25] and gastric hypomotility is documented in Duchenne dystrophy.[26]

RESPIRATORY PROBLEMS

The respiratory complications which inevitably accompany the later stages of the neuromuscular diseases[27-29] are directly due to inspiratory muscle weakness and to an ineffective cough mechanism due to expiratory muscle weakness. Clearly it is important to discourage patients from smoking. It is essential for family members to learn to do postural drainage and assisted coughing promptly in the event of a chest infection.[7,9,10]

A newly recognized problem in advanced cases of Duchenne muscular dystrophy is a tendency to brief interruptions in breathing during the 'rapid eye movement' (REM) sleep when dreaming takes place, which if prolonged can result in oxygen lack.[30,31]

Fortunately the respiratory problems in the muscular dystrophies is now a matter of interest to the European Alliance of Muscular Dystrophy Associations which has set up a specialist Working Party which has made recommendations[32,33] for providing respiratory support at night to those patients who need it. In practice benefit has been gained from the use of cuirass type respirators or nasal intubation with intermittent positive pressure ventilation in selected patients with Duchenne, Becker and limb girdle dystrophies. A positive approach to respiratory support is also well pursued in some European countries and in the USA.[27,29,34,35] It remains to be seen whether this has a significant effect on life expectancy. Despite reports of some patients who appear to have benefited greatly, life expectancy in Duchenne muscular dystrophy has changed little on the average (despite improvements in the management of respiratory infections with antibiotics) over the last century.[36] It also remains to be proved whether the use of internal stabilization of the spine[12,21,22] to avoid the serious respiratory consequences of uncontrolled spinal curvature[30,37] helps prolong life. In 'How briefly my son'[38] Joan Neville described the last illness of her son with Duchenne muscular dystrophy who died of pneumonia shortly before he would have become 13 years of age. It is certainly hoped that more Duchenne patients can now reach their teens and even survive into adulthood, but a policy of respiratory support has not been implemented for sufficiently long or consistently enough to see an effect on average mortality rates. It is clearly important to seek clear confirmation of what is believed to be the pre-eminent importance of maintaining adequate oxygenation.

Respiratory failure also complicates the later stages of the other dystrophies such as the Becker and myotonic types. In the latter there are also sleep-related[39] disturbances of oxygenation, and benefit can follow the use of the drug Mazindol which appears to reduce the duration (and number) of the central apneas in selected patients with the result that the patient feels less somnolent and more alert the next day.[40]

CARDIAC PROBLEMS

Cardiac abnormalities are common in Duchenne muscular dystrophy and changes can readily be detected by noninvasive means.[41,42] These changes are less commonly seen in the adult-presenting dystrophies but they should always be sought by the relevant investigations. In myotonic dystrophy conduction defects are common[25] and these may require pacing. Cardiac failure is fairly unusual although 'Cor Pulmonale' resulting from chronic respiratory failure is not unknown. One factor which may explain the lack of obvious cardiac failure in teenagers with late Duchenne or in adults with muscular dystrophy who have cardiac involvement is the immobilization which keeps demands on the heart to a minimum. The cardiac management can usually be sustained with standard drug therapy. When cardiac failure appears intractable it is important to consider whether respiratory support could help reverse the effects of hypoxia on the heart. Cardiac problems may become a more prominent feature of advanced muscular dystrophy when a more positive attitude to respiratory problems is pursued.

PSYCHOLOGICAL MANAGEMENT

The psychological aspects of the muscular dystrophies are enormous (Table 2). There are several important 'milestones' in the course of the disease which require sympathetic understanding and guidance,[43,44] not least the approach of the inevitable final illness.[38,45]

SOME PSYCHOLOGICAL 'MILE STONES'

Suspicion

The adult-onset muscular dystrophies may appear first as a 'slowing up' of climbing stairs or walking, together with a tendency to trip and fall.

Diagnosis

The confirmation of the patient's suspicion by a specialist may result in a range of confused emotions. It is important to work through these emotions by articulating fears, resentments and worries to a sympathetic person who can explain in an honest but positive way what lies ahead so that the patient achieves a 'coping strategy'.

Loss of walking capability

In Duchenne muscular dystrophy ambulation is lost before the teens (otherwise there is doubt as to the correctness of the diagnosis). If there is an additional factor such a severe intercurrent illness, limb fracture or overweight, ambulation is likely to be lost earlier whether the dystrophy is Duchenne or one of the adult types. Once the patient becomes very dependent on others for transfers, the key advice is to ask how he or she wishes to be moved or positioned. It is as well to observe how the family is used to carrying out transfers or other movements. This is important not only to avoid unnecessary discomfort but also to avoid risk of injury or embarassment.

However disabled the patient by weakness and its complications the spirit of personal survival is indomitable and it is impressive to note (not infrequently) the artistic talents of some patients (e.g. Fig. 2).

Fig. 2 Pen and ink drawing of a bird by a young adult patient severely disabled by facioscapulohumeral muscular dystrophy complicated by paraplegia. The drawing was presented to the author by the patient who kindly gave permission for it to be reproduced here.

Loss of driving licence

Here the medical attendant seeks to sustain mobility by car as long as possible but it must be emphasized that on each visit a verbal check is made as to whether the patient would react in an emergency in a 'fail dangerous' way. By this it is meant that in the event of a swerve or the need to brake suddenly would the patient be thrown off balance and lose control completely? Such considerations should also be extended to motorway driving at speed and whether the patient insists on driving the whole family. If there is any doubt as to the patient's ability to drive safely then a special driving test for the disabled should be insisted upon. This could give advice on what modifications might be needed to the vehicle or whether the driving licence should be terminated.

Loss of employment

Adults with even severe disabilities and handicap due to muscular dystrophies still manage to be in full time employment. Examples are office work, hairdressing, medical research and business. The sickness absence record of such disabled persons is generally satisfactory[46] and patients should not be allowed to feel pressurized into premature retirement on medical grounds just to suit the employers. The 'when' is not always such a problem; it becomes clear to the patient and the family when the struggle to get to work and cope with the work situation is too much.

Sexual relations

Disability puts the patient at an obvious handicap. Only recently have the psychosexual implications of neuromuscular diseases been explored in adult patients.[47]

The later years

For the teenage patient with Duchenne muscular dystrophy it is distressing to see his contemporaries deteriorating and eventually dying while he becomes aware of his own progressive weakness and further disability. So far little has been done formally to explore the attitudes of the patient to his impending demise. My own discussions with individuals who have approached their end with more or less equanimity indicates a progressive weariness with the struggles of life. There may be respiratory distress associated with respiratory muscle weakness or fatigue. Cardiac failure which may dominate the final illness in a few patients is liable to cause dyspnoea, orthopnoea and abdominal discomfort due to hepatic congestion. To realize that the end may come as a sleep without waking may be a source of comfort to the patient and family. Bereavement following the death of a sufferer is not addressed to any extent as yet but there is comfort to be had from reading books which give positive counselling encapsulated in their titles: 'The courage to grieve'[48] and 'When bad things happen to good people'.[49]

HOPE OF EFFECTIVE DRUG TREATMENT

Expectations for effective drug treatment are high because of the discovery of the defective gene locus on the short arm of the X-chromosome in Duchenne muscular dystrophy[50] and the more recent discovery of the gene product, the sarcolemmal cytoskeletal protein 'dystrophin'.[51] This is believed to be important in protecting against mechanical stresses.[52] Such mechanically-induced lesions of the sarcolemma with secondary (calcium-induced) damage to the contractile machinery of muscle are suggested (Fig. 3) as an explanation for the proximal distribution of most human myopathies.[53] The precise genetic defects in the adult-presenting muscular dystrophies are yet to be discovered though the defect in myotonic dystrophy is believed[24] to be carried on Chromosome 19. In this condition it is noteworthy that treatment of the symptom of myotonia (e.g. with phenytoin) does not influence the course of the dystrophy.[25]

The gulf between these exciting discoveries and effective treatment is however so wide that endeavours must continue to seek means of altering the dystrophic processes. Many agents have been tried in a range of dystrophies often with an initially encouraging result but with later disappointment when the drug was subsequently submitted to more rigorous (double blind controlled clinical trial) assessment.[18] While the statistical basis of therapeutic trials in Duchenne dystrophy has been clearly worked out,[54] therapeutic trials in the adult-presenting muscular dystrophies are more difficult because of the great variability and chronicity of progress.

In attempts to determine the pharmacology of possible therapeutic agents we have particularly studied the effects on muscle

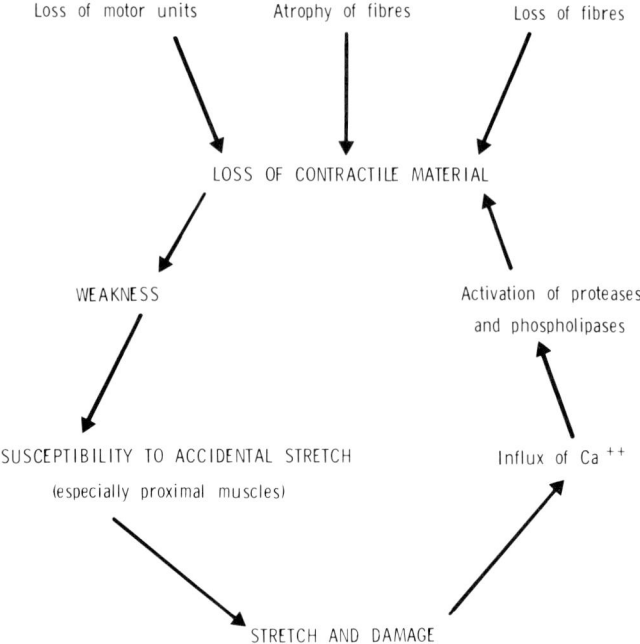

Fig. 3 Hypothetical vicious circle[53] of mechanically-induced damaged to skeletal muscle which could be contributing to disease progress in adults with muscular dystrophy.

metabolism. Agents so far investigated influence muscle protein synthesis rate which was found to be reduced in teenagers with Duchenne muscular dystrophy,[55] and adults with myotonic dystrophy[56] but these were not found to give clinical benefit despite increasing muscle protein synthesis rate. Agents presumed to influence free radical protection (e.g. Vitamin E) have also been investigated but without convincing evidence of beneficial change in muscle metabolism.[57] Parallel laboratory-based pharmacological studies in animal models suggest agents which may reduce secondary calcium-induced muscle damage[58] but no trials have been made in patients.

The muscular dystrophies and possible drug treatments come under the category of 'Orphan Diseases and Orphan Drugs'.[59] The responsibility for further research rests on the interested parties (patients, families, charities and relevant medical scientists) rather than on the pharmaceutical companies who can expect little profit from the effective treatment of these rare diseases. Nevertheless there are grounds for optimism that an integrated approach to management (including institutional care[60]) such as that described can do much to improve the quality and possibly the duration of life in adult muscular dystrophy patients.

ACKNOWLEDGEMENTS

The support of the Muscular Dystrophy Group of Great Britain and Northern Ireland is gratefully acknowledged for clinics and research programme which provide the experience described in this paper.

REFERENCES

1 Edwards RHT, Griffiths RD, Hayward M, Helliwell, TR. Modern methods of diagnosis of muscle diseases. J R Coll Physicians Lond 1986; 20: 49–55
2 Walton JN, Gardner-Medwin D. The Muscular Dystrophies. In: Walton JN, ed. Disorders of Voluntary Muscle (5th edn). Edinburgh: Churchill Livingstone, 1988: pp. 519–568
3 Edwards RHT, Round JM, Jones DA. Needle biopsy of skeletal muscle: a review of 10 years' experience. Muscle Nerve 1983; 6: 676–683
4 Dietrichson P, Coakley J, Smith PEM, Griffiths RD, Helliwell TR, Edwards RHT. Concnotome and needle percutaneous biopsy of skeletal muscle. J Neurol Neurosurg Psychiatr 1987; 50: 1461–1467
5 Siegel IM. Everybody's different, nobody's perfect. Muscular Dystrophy Association America/Muscular Dystrophy Group of Great Britain, 1982; 1–13
6 The Muscular Dystrophy Handbook: A practical Guide for those who suffer from muscular dystrophy and allied neuromuscular diseases (New Edition). London: The Muscular Dystrophy Group of Great Britain and Northern Ireland
7 Edwards RHT. Advice to Patients On Care Of Respiratory Function In Muscular Dystrophy. London: The Muscular Dystrophy Group of Great Britain and Northern Ireland
8 Edwards RHT. Weight Control on Patients with Muscular Dystrophy. London: The Muscular Dystrophy Group of Great Britain and Northern Ireland
9 Chapman S, Newham D. General Exercise. In: A Guide to Exercise for the adult patient with Muscular Dystrophy. London: The Muscular Dystrophy Group of Great Britain and Northern Ireland
10 Hyde SA. The parents' guide to physical management of Duchenne Muscular Dystrophy. London: The Muscular Dystrophy Group of Great Britain and Northern Ireland
11 Harpin P. With a little help: A guide to aids and adaptations for people with muscular dystrophy and allied neuromuscular diseases. London: The Muscular Dystrophy Group of Great Britain, vol. I–VIII
12 Siegel IM. Muscle and its diseases. Chicago: Year Book Medical Publishers, 1986
13 Khodadadeh S, McClelland MR, Patrick JH, Edwards RHT, Evans GA. Knee moments in Duchenne Muscular Dystrophy. Lancet 1986; ii: 544–545
14 Edwards RHT, Round JM, Jackson MJ, Griffiths RD, Lilburn MF. Weight reduction in boys with muscular dystrophy. Dev Med Child Neurol 1984a; 26: 375–383
15 Griffiths RD, Edwards RHT. A new chart for weight control in Duchenne muscular dystrophy. Arch Dis Child 1988; 63: 1256 1258

16 Griffiths RD. Controlling weight in muscle disease to reduce burden. Physiotherapy 1988; (In press)
17 Vignos P. Exercise in neuromuscular disease: statement of the problem. In: Neuromuscular Diseases. Serratrice G, et al. eds. New York: Raven Press, 1984: pp. 565–569
18 Dubowitz V, Heckmatt J. Management of muscular dystrophy. Pharmacological and physical aspects. Br Med Bull 1980; 36: 139–144
19 Burford K, Pentland B. Management of Motor Neurone Disease: the Physiotherapist's role. Physiotherapy 1985; 71: 402–404
20 Norris FH, Smith RA, Denys EH. Motor neurone disease: towards better care. Br Med J 1985; 291: 259–262
21 Rideau Y, Glorion B, Delaubier A, Tarle O, Bach J. The treatment of scoliosis in Duchenne muscular dystrophy. Muscle Nerve 1984; 7: 281–286
22 Sussman MD. Advantage of early spinal stabilization and fusion in patients with Duchenne muscular dystrophy. J Paediatr Orthopaed 1984; 4: 532–537
23 Galasko CSB. The orthopaedic management of the dystrophies, myopathies, atrophies, neuropathies and ataxias. In Neuromuscular problems in orthopaedics. Oxford: Blackwell, 19??: pp. 83–105
24 Harper PS. The myotonic disorders. Chapter 16 In: Walton JN ed. Disorders of Voluntary Muscle (5th edn). Edinburgh: Churchill Livingstone, 1988; pp. 569–587
25 Harper PS. Myotonic Dystrophy. Philadelphia: Saunders, 1979
26 Barohn RJ, Levine EJ, Olsen JO, Mendell JR. Gastric hypomotility in Duchenne's muscular dystrophy. N Engl J Med 1988; 319: 15–18
27 Rideau Y, Jankowski LW, Grellet J. Respiratory Function in the muscular dystrophies. Muscle Nerve 1981; 4: 155–164
28 Nigro G, Comi L, Limongelli FM, Giugliano MAM et al. Prospective study of X-linked progressive muscular dystrophy in Campania. Muscle Nerve 1983; 6: 253–262
29 Ringel SP, Martin RJ, Libby LS. Newer treatments of respiratory dysfunction in neuromuscular diseases. In: Carraroll, Angelini C, eds. Cell Biology and Clinical Management in Functional Electrostimulation of Neurones and Muscles. Cleup Editore Padova 1985; 217–221
30 Smith PEM, Calverley PMA, Edwards RHT et al. Practical Problems in the Respiratory Care of Patients with Muscular Dystrophy. N Engl J Med 1987; 316: 1197–1205
31 Smith PEM, Calverley PMA, Edwards RHT. Hypoxia during sleep in Duchenne muscular dystrophy. Am Rev Respir Dis 1988; 137: 884–888
32 European Alliance of Muscular Dystrophy Associations (EAMDA) Report. Respiratory Insufficiency. Vereniging Spierziekten Nederland. Lt Gen van Heutszlaan 6, 3743 JN Baarn, The Netherlands, 1987
33 Spencer, GT. Domestic respiratory care. European Alliance of Muscular Dystrophy Associations (EAMDA): Proceedings of 17th European Conference, Oslo September 1987. Published by Foreningen for Muskelskye. Sporveisgata 10, N-0354 Oslo 3, Norway 1988; pp. 20–22
34 Raphael JC. Ethical aspects of respiratory care. Observation from trial studies in France. European Alliance of Muscular Dystrophy Associations (EAMDA): Proceedings of 17th European Conference, Oslo September 1987. Published by Foreningen for Muskelskye. Sporveisgata 10, N-0354 Oslo 3, Norway 1988; pp. 33–35
35 Curran FJ. Night Ventilation by Body Respirators for Patients in Chronic Respiratory Failure Due to late Stage Duchenne Muscular Dystrophy. Arch Phys Med Rehabil 1981; 36: 139–144
36 Emery A. Duchenne muscular dystrophy. Oxford: Oxford University Press, 1986

37 Kurtz LT, Mubarak SJ, Schultz P, Park SM, Leach J. Correlation of scoliosis and pulmonary function in Duchenne muscular dystrophy. J Pediatr Orthoped 1983; 3: 347–353.
38 Neville J. So briefly my son. London: Hutchinson, 1962; pp. 1–80
39 Cirignotta F, Mondini S, Zucconi M et al. Sleep related breathing impairment in myotonic dystrophy. J Neurol 1987; 235: 80–85
40 Coakley JH, Edwards RHT, Calverley PMA. The effect of Mazindol on central apnoeas in myotonic dystrophy. Eur J Clin Invest (In preparation)
41 Comi LI, Salvatore M, Limongelli FM et al. Cardiomyopathies in X-linked progressive muscular dystrophy. Evaluation with noninvasive methods. Cardiomyology 1982; 1: 79–90
42 Nigro G. Cardiac Insufficiency. European Alliance of Muscular Dystrophy Associations (EAMDA): Proceedings of 17th European Conference, Oslo September 1987. Published by Foreningen for Muskelsyke. Sporveisgata 10, N-0354 Oslo 3, Norway 1988. p. 36. (Introducing EAMDA White Paper 'Cardiac Insufficiency'—Guidelines of impending cardiac failure in patients suffering from muscular and neuromuscular diseases 1988; pp. 1–20
43 Edwards RHT. Medical and Psychological Management of Neuromuscular Disease. In: Walton JN ed. Disorders of Voluntary Muscle (5th edn). Edinburgh: Churchill Livingstone, 1988: pp. 719–729
44 Ringel SP. Neuromuscular Disorders. A guide for patients and family. New York. Raven Press, 1987
45 Madorsky JGB, Radford LM, Newman EM. Psychological Aspects of Death and Dying in Duchenne Muscular Dystrophy. Arch Phys Med Rehabil 1984; 65: 79–82
46 Edwards FC, McCallum RI, Taylor PJ, eds. Fitness for work. Report of the Royal College of Physicians. Oxford Medical Publications, 1988
47 Nordqvist I. Sexuality and human relationships. The role of attitudes and solutions in sexual life for individuals with neuromuscular disorders. European Alliance of Muscular Dystrophy Associations (EAMDA): Proceedings of 17th European Conference, Oslo September 1987. Published by Foreningen for Muskelsyke. Sporveisgata 10, N-0354 Oslo 3, Norway 1988; pp. 16–19
48 Tatelbaum J. The courage to grieve. London: Heinemann; 1980
49 Kushner HS. When bad things happen to good people. London: Pan Books, 1981, pp. 54–79
50 Monaco AP, Neve RL, Colletti-Feener CA, Bartelson CJ, Kurnitt DM and Kunkel LM. Isolation of candidate cDNAs for portions of the Duchenne Muscular Dystrophy gene. Nature 1985; 323: 646–650
51 Hoffman EP, Fischbeck KH, Brown RH Jr et al. Dystrophin characterisation in muscle biopsies from Duchenne and Becker muscular dystrophy patients. N Engl J Med 1988; 318: 1363–1368
52 Karpati G, Carpenter S. The deficiency of a sarcolemmal cytoskeletal protein (dystrophin) leads to the necrosis of skeletal muscle fibers in Duchenne/Becker dystrophy. In: Proceedings of the Eric K Fernstrom Foundation Symposium 'The Neuromuscular Junction' Lund, Sweden June 1988. Amsterdam: Elsevier (In Press)
53 Edwards RHT, Newham DJ, Jones DA, Chapman SJ. Role of mechanical damage in pathogenesis of proximal myopathy in man. Lancet 1984; i: 548–552
54 Brooke MH, Fenichel GM, Griggs RC et al. Clinical investigation in Duchenne muscular dystrophy: 2 Determination of the 'power' of therapeutic trials based on the natural history. Muscle Nerve 1983; 6: 91–103
55 Rennie MJ, Edwards RHT, Millward DJ, Wolman SL, Halliday D, Matthews DE. Effects of Duchenne muscular dystrophy on muscle protein synthesis. Nature 1982; 296: 165–167
56 Halliday D, Ford GC, Edwards RHT, Rennie MJ, Griggs RC. In vivo

estimation of muscle protein synthesis in myotonic dystrophy. Ann Neurol 1985; 17: 65–69

57 Edwards RHT, Jones DA, Jackson MJ. An approach to treatment trials in muscular dystrophy with particular reference to agents influencing free radical damage. Med Biol 1984; 62: 143–147

58 Jackson MJ, Jones DA, Edwards RHT. Vitamin E and Muscle Diseases. J Inherit Metab Dis 1985b; 8 (Suppl 1): 84–87

59 Herbert I, Scheinberg, Walshe JM (eds). Orphan Diseases and Orphan Drugs. Published by Fulbright Papers, Manchester University Press, 1985

60 Report of the Royal College of Physicians: The Young Disabled Adult. London: Royal College of Physicians, Sept 1986.

British Medical Bulletin (1989) Vol. 45, No. 3, pp. 819–824

Future prospects

P J Lachmann
MRC Molecular Immunopathology Unit, MRC Centre, Cambridge, UK

In the last few years the biochemical lesion in
Duchenne/Becker Muscular Dystrophy has been
discovered and the gene for the missing/abnormal protein
(dystrophin) has been cloned.
The implications of these discoveries on the prospects
for diagnosing and eventually eradicating this disease and
for the possible treatment of affected boys are discussed.

The last few years have seen critical progress in our understanding
of the molecular basis of the Duchenne/Becker form of muscular
dystrophy and this number of the British Medical Bulletin is an
indicator of how far and how fast this progress has gone. This
disease has provided the first major triumph of the techniques of
'reverse genetics' where the nature of an abnormal phenotype is
determined from studies of the affected genome rather than vice
versa. It is now clear that the disease is caused by the absence (or
presence in an abnormal form) of a hitherto undescribed muscle
protein which has been given the name of dystrophin and whose
exact physiological functions remain to be determined. The gene
is exceptionally large, which may account for the frequency of new
mutation, and there is a high incidence of gene deletions. The
genetic lesions within the dystrophin gene are highly variable as is
the case in a number of other genetic diseases, notably the
thalassemias.

Armed with this new knowledge what are the prospects now for
the eradication of Duchenne muscular dystrophy (DMD) and for
the possible treatment of those who are already affected.

PREVENTION OF THE DISEASE

It is probably already possible with the currently available gene
probes to diagnose the great majority, if not quite all, affected boys
by DNA analysis so that chorion villus biopsy should allow early

0007–1420/89/0045–0819/$10.00

antenatal diagnosis. Where there has been a previous affected boy within a family there is every reason to be optimistic that the prevention of further cases will be almost totally efficient. Where any doubt remains, a fetal muscle biopsy can be tested for dystrophin—its absence being a sure marker of DMD. This will, however, prevent only about one half to two thirds of new cases since the remainder occur in families not previously known to be at risk and where a new mutation has occurred either in the carrier mother (or in one of her female ancestors) or in the fetus himself. There is some argument whether the frequency of the new mutation occurring in the mother and the fetus are equal or whether more occur in the mother. This is of importance since it is much easier to screen DNA from pregnant women than it is to undertake chorion villus biopsies on all fetuses. It seems likely that in the not too distant future all pregnant women will be offered DNA analysis to determine carrier status not only for DMD but also for other genetic diseases for which probes are available. A list of genetic diseases where this approach is likely to be useful is shown in Table 1. It is difficult to give an estimate of cost-effectiveness for such a procedure but it is likely to be high since the cost of looking after affected children not only with muscular dystrophy but with cystic fibrosis and above all perhaps with mental deficiency is so high both in financial and human terms. Furthermore, there are no real ethical problems involved since women who would not wish either to have advice on mate selection (in the case of autosomal recessive diseases) or to accept early termination in the event of having an affected fetus would simply not have the investigation done.

A much more radical innovation would be the introduction of universal chorion villus biopsy for the diagnosis of serious genetic disease. The cost of such a procedure would be much higher since it requires an invasive procedure carried out by an expert and, though in good hands chorion villus biopsy appears to be very safe, it will undoubtedly carry some complication rate. It will on the other hand be more efficient at preventing genetic disease since it is the only way of picking up mutations that occur in the fetus itself. It is also much easier to pick up hemizygotes or homozygotes for a mutation than it is to pick up heterozygotes and the diagnostic technique is therefore likely to be more reliable. It is difficult to believe that such an approach will be adopted on a large scale in the foreseeable future and certainly not until all the major common genetic diseases can be confidently diagnosed at the DNA

Table 1   Some conditions diagnosable by DNA probes

Adrenoleukodystrophy
α_1-antitrypsin deficiency
Adrenal hyperplasia (Congenital AH; 21-OH deficiency)
Chromosomal (trisomies etc)
Coagulation Factor VIII, IX and XI deficiency
Cystic fibrosis
Friedreich's ataxia
Haemoglobinopathies
Huntington's disease
Hypoparathyroidism
Immunodeficiency (X-linked)
Myotonic dystrophy
Ornithine transcarbamylase deficiency
Phenylketonuria
Polycystic kidney disease (adult)
Retinitis pigmentosa
Retinoblastoma
von Willebrand's disease
X-linked muscular dystrophies (Duchenne/Becker)
X-linked mental deficiency (not yet efficient)

level. It is, however, likely that experience will be gained with groups of pregnancies, particularly those affecting older women, where the risk of chromosomal abnormalities in the fetus is much higher and where there is therefore other justification for chorion villus biopsy. Preliminary studies of this kind will give an indication of how cost effective the universal approach might in the long term be.

THE TREATMENT OF DUCHENNE/BECKER MUSCULAR DYSTROPHY

The treatment of genetic diseases is seldom easy. Up to the present time the only possible approach was to attempt to circumvent the genetic lesion in some way, there being no way of correcting the genetic lesion itself. Occasionally this was quite successful. For example in phenylketonuria where there is absence of the enzyme that oxidises phenylalanine to tyrosine the harmful effects on the brain giving rise to mental deficiency were found to be due to the presence of toxic levels of phenylalanine derived from proteins in the diet especially early in life. Diagnosis by testing of blood obtained by a heel prick at birth and putting affected babies on a diet very low in phenylalanine for the first years of their life has had a striking effect on their prognosis. Similarly the level of the toxic metabolites that give rise to immunity deficiency in adenine deaminase deficiency can be reduced by transfusions of normal red cells which contain enough enzyme to clear the plasma of much of the toxic material. However such treatments can be devised only if the nature of the biochemical defect and the way in which it causes the observed damage are clearly understood. Until recently neither has been the case for the Duchenne/Becker muscular dystrophy. Now that it is known that the basic defect is the absence of dystrophin or the production of a dysfunctional molecule there is intensive investigation of the function of dystrophin in normal muscle; but the link between the absence of the protein and the appearance of the slowly developing muscle damage that characterizes this muscular dystrophy is not yet fully understood. Devising a rational treatment is therefore not so far possible. However some suggestions have been made. One arises from the fact that the mice that are deficient in dystrophin produce only a trivial clinical dystrophy and their muscles do not show fibrosis. It has therefore been suggested that in man it may be the fibrosis in muscle, possibly subsequent upon repeated cycles of some form of biochemical insult and repair, that eventually does the real damage. Attempts to prevent the fibrotic process are not easy although drugs that inhibit inflammation such as corticosteroids may help to do this in certain circumstances. However experimentation with the removal of fibrous tissues from the fascia surrounding muscles of the leg is now being tried and the original, uncontrolled and anecdotal accounts sound promising. A randomised trial is now being initiated under the auspices of the Muscular Dystrophy Group so it will be possible to see if the removal of fibrous tissue from children at a relatively early stage in their disease improves their prognosis.

The second line of approach is to see whether the genetic lesion can be modified by external stimuli. This will not be possible if there is deletion of the whole gene or of sufficiently large parts of it that no useful protein can be made from it. However, if the genetic lesion were to be a non-coding region so that transcription fails then it may be possible by giving stimulants of transcription—androgenic hormones or gamma interferon for example—to increase the rate of production. In a human disease which is not at all analogous since the lesion is heterozygous—the deficiency of C1 inhibitor which causes hereditary angio-oedema—it is now well established that treating patients with modified androgenic hormones, such as danazol or stanazalol, can increase the rate of production from the normal gene sufficiently to 'cure' the disease.

It is likely that the majority of children with muscular dystrophy have deleted exons. It is believed that in Duchenne dystrophy the deletions are 'out of frame' so that the protein is truncated to the beginning of the deletion if it is made at all; while in Becker dystrophy the deletions may be 'in phase' so that a protein just missing the deleted region may be made and it is not clear that interfering with transcription and translation will improve this situation. Some apparent deficiencies in other systems, however, are due to changes in the protein which render it susceptible to intracellular proteolysis. This is believed to be the basis of C3 deficiency in guinea pigs[1] but I am not aware that this has so far been successfully treated. However interference with the activity of intracellular proteolytic enzymes is not beyond the realm of possibility.

However, the real long-term hope for treating genetic diseases is to restore the defect in the genome with a normal gene and it is to this prospect that the term 'gene therapy' is usually applied. In principle, gene therapy is now possible. Genes can be inserted into retroviral vectors which can infect mammalian somatic cells and expression, even in vivo, has been demonstrated.[2,3] So far it has, however, proved extremely difficult to produce long-lasting, well controlled and high level expression or indeed to ensure the infection of a sufficiently large number of cells and these are all problems that will require to be solved before gene therapy in humans becomes a practical proposition. It seems likely that it will be necessary (or at least extremely helpful) to have a selective system which allows the infected cells a growth advantage over uninfected cells. For this reason alone one of the first diseases that is being investigated in this respect is Lesch-Nyhan syndrome due to deficiency of the enzyme, hypoxanthine guanine phosphoribosyl transferase (HGPRT). Selection based on this enzyme deficiency is widely used in making monoclonal antibodies. Myeloma cells deficient in HGPRT are selected by growing the cells in 8-azaguanine (which kills normal cells but not HGPRT deficient cells). The hybridomas, made by fusion of the deficient myeloma cells with normal spleen cells, are then selected by growth in a medium containing hypoxanthine, aminopterin and thymine (HAT) which kills HGPRT deficient cells selectively. In vitro therefore, bone marrow cells from a Lesch-Nyhan child successfully infected with a retrovirus bearing HGPRT can be selected in HAT medium and grown to increase their number. They may then be reinfused into the patient. In the case of muscle, the situation is complicated by the syncitial nature of striated muscle fibres. Conceivably, the introduction of a proportion of modified nuclei into muscle would correct a genetic defect. There are, however, particular difficulties to cope with in trying to restore the dystrophin gene. The gene is very large, even the coding region extending over 16kb. This is too big for the existing retroviral vectors—and if the control regions of the gene are to be introduced as well this problem will be more severe. There are other ways of introducing DNA into cells in vitro not necessarily using a vector, but their efficiency is much lower. Even so and notwithstanding the striking speed at which molecular biology is progressing, gene therapy for muscular dystrophy does not seem to be imminent and the more immediate hopes must be focused on the other forms of therapy here described.

REFERENCES

1 Auerbach HS, Burger R, Bitter-Suermann D, Goldberger G, Colten HR. C3-deficient guinea pig mRNA directs synthesis of a structurally abnormal C3 protein. XIth Int Complement Workshop, Miami Fl, 1985. Abstract. Complement 1985; 2: 5.

2 Mann R, Mulligan RC, Baltimore DB. Construction of a retrovirus packaging mutant and its use to produce helper free defective retrovirus. Cell 1983; 33: 153–159.

3 Dzierzak EA, Papayannopoulou T, Mulligan RC. Lineage-specific expression of a human γ-globin gene in murine bone marrow transplant recipients reconstituted with retrovirus-transduced stem cells. Nature 1988; 331: 35–41.

Index